AMERICAN POLITICAL, ECONOMIC, AND SECURITY ISSUES SERIES

THE JOINT OPERATING ENVIRONMENT 2008: USING HISTORY TO INFORM THE FUTURE

AMERICAN POLITICAL, ECONOMIC, AND SECURITY ISSUES SERIES

The Land and Maritime Boundary Disputes of the Americas
Rongxing Guo
2009. ISBN: 978-1-60741-636-4

The Call for A New Era in the U.S. Budget
Benjamin J. Walters (Editor)
2010. ISBN: 978-1-60741-883-2

Closing Guantanamo - Issues and Legal Matters Surrounding the Detention Centers End
Noah M. Claeys
2010. ISBN: 978-1-60741-511-4

Federal Spending: A Riddle Wrapped up in an Enigma
Frederick C. Hargis
2010. ISBN: 978-1-60741-728-6

Foreign Investment In U.S. Securities
Declan J. Byrne (Editor)
2010. ISBN: 978-1-60876-084-8

The State Secrets Privilege and Other Limits on Classified Information
Jonathon W. Collingsworth (Editor)
2010. ISBN: 978-1-60876-967-4

The State Secrets Privilege and Other Limits on Classified Information
Jonathon W. Collingsworth (Editor)
2010. ISBN: 978-1-60876-430-3 (Online)

The Joint Operating Environment 2008: Using History to Inform the Future
Michaël Peeters (Editor)
2010. ISBN: 978-1-60876-035-0

AMERICAN POLITICAL, ECONOMIC, AND SECURITY ISSUES SERIES

THE JOINT OPERATING ENVIRONMENT 2008: USING HISTORY TO INFORM THE FUTURE

MICHAËL PEETERS
EDITOR

Nova Science Publishers, Inc.
New York

Copyright © 2010 by Nova Science Publishers, Inc.

All rights reserved. No part of this book may be reproduced, stored in a retrieval system or transmitted in any form or by any means: electronic, electrostatic, magnetic, tape, mechanical photocopying, recording or otherwise without the written permission of the Publisher.

For permission to use material from this book please contact us:
Telephone 631-231-7269; Fax 631-231-8175
Web Site: http://www.novapublishers.com

NOTICE TO THE READER

The Publisher has taken reasonable care in the preparation of this book, but makes no expressed or implied warranty of any kind and assumes no responsibility for any errors or omissions. No liability is assumed for incidental or consequential damages in connection with or arising out of information contained in this book. The Publisher shall not be liable for any special, consequential, or exemplary damages resulting, in whole or in part, from the readers' use of, or reliance upon, this material. Any parts of this book based on government reports are so indicated and copyright is claimed for those parts to the extent applicable to compilations of such works.

Independent verification should be sought for any data, advice or recommendations contained in this book. In addition, no responsibility is assumed by the publisher for any injury and/or damage to persons or property arising from any methods, products, instructions, ideas or otherwise contained in this publication.

This publication is designed to provide accurate and authoritative information with regard to the subject matter covered herein. It is sold with the clear understanding that the Publisher is not engaged in rendering legal or any other professional services. If legal or any other expert assistance is required, the services of a competent person should be sought. FROM A DECLARATION OF PARTICIPANTS JOINTLY ADOPTED BY A COMMITTEE OF THE AMERICAN BAR ASSOCIATION AND A COMMITTEE OF PUBLISHERS.

LIBRARY OF CONGRESS CATALOGING-IN-PUBLICATION DATA

The joint operating environment 2008 : using history to inform the future / Michaël Peeters, editor .
   p. cm.
  Includes bibliographical references and index.
  ISBN 978-1-60876-035-0 (softcover : alk. paper)
  1. Unified operations (Military science). 2. Combined operations (Military science). 3. National security. 4. Security, International. I. Peeters, Michaël.
  U260.J655 2010
  355.4'6--dc22
                                                                                    2009038416

Published by Nova Science Publishers, Inc. ✛ New York

CONTENTS

Preface vii

Chapter 1 U.S. Joint Forces Command Clarifies Joint Operating Environment 2008   1
Michaël Peeters

Chapter 2 The Joint Operating Environment 2008: Challenges and Implications for the Future Joint Force   3
United States Joint Forces Command

Index 91

PREFACE

The next quarter century will challenge U.S. joint forces with threats and opportunities ranging from regular and irregular wars in remote lands, to relief and reconstruction in crisis zones, to sustained engagement in the global commons. It is impossible to predict precisely how challenges will emerge and what form they might take. Nevertheless, it is absolutely vital to try to frame the strategic and operational contexts of the future, in order to glimpse at the possible environments where political and military leaders will work and where they might deploy joint forces. Furthermore, the challenges of the future will resemble, in many ways, the challenges that American forces have faced over the past two centuries. War has been a principal driver of change over the course of history and there is no reason to believe that the future will differ in respect. In contrast, changes in the strategic landscape, the introduction and employment of new technologies, and the adaptation and creativity of our adversaries will alter the character of joint operations a great deal. The authors of this book reason that only through reading the signposts of the times will the Joint Force have some of the answers to the challenges of the future. This book consists of public documents which have been located, gathered, combined, reformatted, and enhanced with a subject index, selectively edited and bound to provide easy access.

This is an edited, excerpted and augmented edition of a United States Joint Forces Command, Background Paper, Joint Operating Environment publication, dated November 25, 2008.

In: The Joint Operating Environment 2008: ISBN: 978-1-60876-035-0
Editors: Michaël Peeters © 2010 Nova Science Publishers, Inc.

Chapter 1

U.S. JOINT FORCES COMMAND CLARIFIES JOINT OPERATING ENVIRONMENT 2008

Michaël Peeters

U.S. Joint Forces Command releases a clarification on the Joint Operating Environment 2008.

(NORFOLK, Va. – Dec. 10, 2008) -- On Dec. 4, 2008, U.S. Joint Forces Command (USJFCOM) released the Joint Operating Environment 2008 (JOE 2008), a report that discusses the trends and contexts of the future operating environment and their implication for the future joint force.

JOE 2008 is designed to spark discussions with national security and multinational partners about the nature of the future security environment and its potential implications for the future joint force.

The JOE 2008 contains a statement regarding nuclear powers on page 32. The statement regarding North Korea does not reflect official U.S. government policy regarding the status of North Korea. The U.S. government has long said that we will never accept North Korea as a nuclear power. This clarification has been communicated to the embassy of the Republic of Korea.

This JOE is fundamentally speculative in nature and is intended to serve as a starting point for discussions about the future security environment. The Joint Operating Environment is not meant to be a statement of policy.

In the broadest sense, the *Joint Operating Environment* examines three questions:
1. What future *trends* and disruptions are likely to affect the joint force over the next quarter century?
2. How are these trends and disruptions likely to define the future *contexts* for joint operations?
3. What are the *implications* of these trends and contexts for the joint force?

By exploring these trends, contexts, and implications, the Joint Operating Environment provides a basis for thinking about the world a quarter of a century from now. Its purpose is not to predict, but to suggest ways leaders might think about the future.

In: The Joint Operating Environment 2008:
Editors: Michaël Peeters

ISBN: 978-1-60876-035-0
© 2010 Nova Science Publishers, Inc.

Chapter 2

THE JOINT OPERATING ENVIRONMENT 2008: CHALLENGES AND IMPLICATIONS FOR THE FUTURE JOINT FORCE

United States Joint Forces Command

FOREWORD

Predictions about the future are always risky. Admittedly, no one has a crystal ball. Regardless, if we do not try to forecast the future, there is no doubt that we will be caught off guard as we strive to protect this experiment in democracy that we call America.

The Joint Operating Environment (JOE) is our historically informed, forward-looking effort to discern most accurately the challenges we will face at the operational level of war, and to determine their inherent implications. We recognize that the future environment will not be precisely the one we describe; however, **we are sufficiently confident of this study's rigor that it can guide future** concept development. While no study can get the future 100 percent correct, we **believe it's most important that we get it sufficiently right, and that the daunting** challenge of perfection not paralyze our best efforts. When future war comes, our concept developers across the Armed Services should have the fewest regrets if today they study, challenge, and implement solutions to the security implications defined here in the JOE. In our line of work, having the fewest regrets defines success when the shocks of conflict bring the surprise that inevitably accompanies warfare.

America retains both the powers of "intimidation and inspiration." We will continue to play a leading role in protecting the values that grew out of the wisdom and vision of our nation's original architects. We must be under no illusions about the threats to our shared values, but we must also recognize the military as only one, albeit critical aspect of America's strength. This strength must specifically recognize the need to adapt to the security challenges we face, whether or not the enemy chooses to fight us in the manner that we would prefer. America's military cannot be dominant yet irrelevant to our policy makers' requirements.

As the JOE goes to print, we face a challenging set of circumstances. The JOE maintains a longer term view and avoids a preclusive vision of future war. Any enemy worth his salt will adapt to target our perceived weaknesses, so the implications contained in this study cannot be rank ordered. But the implications do serve as the basis of the companion Capstone Concept for Joint Operations (CCJO), which outlines how the Joint Force will operate in the future. If the JOE serves as the "problem statement," the CCJO serves as the way the Joint Force will operate in the future to "solve" the problem. These two documents should be seen as two parts of the whole.

In a field of human endeavor as fraught with uncertainty as war, it is essential that we maintain an open mind in our approach. Our responsibility is to turn over this military to our successors in better condition than we who serve today received it. We encourage criticism of our work. We plan to update the JOE routinely in response to your input. Creativity in technological development, operational employment, and conceptual framework is necessary, and it's our intent that the JOE inspires an openness to change so urgently needed when both high- and low-intensity threats abound.

J. N. MATTIS
General, U.S. Marine Corps., Commander, U.S. Joint Forces Command

"War is a matter of vital importance to the State; the province of life or death; the road to survival or ruin. It is mandatory that it be thoroughly studied."[1]

Sun Tzu

INTRODUCTION

The next quarter century will challenge U.S. joint forces with threats and opportunities ranging from regular and irregular wars in remote lands, to relief and reconstruction in crisis zones, to sustained engagement in the global commons. During this time, the causes of conflict will vary from rational political calculation to **uncontrolled passion. Our enemy's capabilities will range** from explosive vests worn by suicide bombers to long-range precision-guided cyber, space, and missile attacks. The threat of mass destruction – from nuclear, biological, and chemical weapons – will likely expand from stable nation-states to less stable states and even non-state networks.

It is impossible to predict precisely how challenges will emerge and what form they might take. Nevertheless, it is absolutely vital to try to frame the strategic and operational contexts of the future, in order to glimpse the possible environments where political and military leaders will work and where they might employ joint forces. The value of such efforts lies not so much in the final product, but much more in the participation of senior leaders and decision-makers in the discussion. Only by wrestling with the possibilities, determining the leading indicators, and then reading the signposts of the times will the Joint Force have some of the answers to the challenges of the future. The alternative, to focus exclusively on the here and now or to pass this mission to the bureaucracy, will certainly result in getting caught flat-footed, reacting to near-term crises as they arise, at great cost in blood and treasure.

Thinking about the future requires an understanding of both what is timeless and what will likely change. As Thucydides suggested in the fifth century BC, **"the events which happened in the past... (human nature being what it is) will at some time or other and in much the same way be repeated in the future."**[2] Many features will not change. The challenges of the future will resemble, in many ways, the challenges that American forces have faced over the past two centuries. In spite of the current intellectual climate in much of the developed world, conflict will not disappear. War has been a principal driver of change over the course of history and there is no reason to believe that the future will differ in this respect. Neither will the fundamental nature of war change. War will remain primarily a human endeavor.

In contrast, changes in the strategic landscape, the introduction and employment of new technologies, and the adaptation and creativity of our adversaries will alter the character of joint operations a great deal. Here too, the past can suggest much about the future – the nature of change, its impact on human societies, and the interplay among human societies in peaceful and warlike competition. While much will stay the same, change will also continue to be a driving force in human affairs.

One cannot rule out the possibility that U.S. military forces will be engaged in persistent conflict over the next quarter century. In the next twenty-five years, there will continue to be those who will hijack and exploit Islam and other beliefs for their own extremist ends. There will continue to be opponents who will try to disrupt the political stability and deny the free access to the global commons that is crucial to the world's economy. In this environment, the presence, reach, and capability of U.S. military forces, working with like-minded partners, will continue to be called on to protect our national interests. Merely sustaining the health of the Joint Force, never mind adapting and transforming, is far more complicated in a period of persistent conflict, with its toll on equipment, people, and national will.

The nature of the human condition will guarantee that uncertainty, ambiguity, and surprise will dominate the course of events. However carefully we think about the future; however thorough our preparations; however coherent and thoughtful our concepts, training, and doctrine; we will be surprised. Even the wisest of statesmen have found their assumptions about the future confounded by reality. The eighteenth century British leader, William Pitt, the Younger, declared in a speech before the House of Commons in February 1792: "Unquestionably there has never been a time in the history of our country when, from the situation in Europe, we might more reasonably expect fifteen years of peace, than we have at the present moment."[3] Within a matter of months, Britain would become embroiled in a conflict that would last nearly a quarter of a century and would kill more Europeans than any other war in history up to that time.

In the broadest sense, the Joint *Operating Environment* examines three questions:

- What future trends and disruptions are likely to affect the Joint Force over the next quarter century?
- How are these trends and disruptions likely to define the future contexts for joint operations?
- What are the implications of these trends and contexts for the Joint Force?

By exploring these trends, contexts, and implications, the *Joint Operating Environment* provides a basis for thinking about the world over the next quarter century. Its purpose is not to predict, but to suggest ways leaders might think about the future.

If war at its essence is a human endeavor, then it follows that one of the most effective ways to understand human nature is by a close consideration of history. As such, rather than futuristic vignettes, the *Joint Operating Environment* uses history as a principal way to gain insight into the future. The discussion begins with the enduring nature of war, the causes and consequences of change and surprises, and the role of strategy. Part II then describes some trends, discontinuities and potential trouble spots that joint forces may confront. Part III analyzes how these trends and disruptions may combine into contexts that will likely define joint operations over the next quarter century. Part IV describes the implications of these contexts for the Joint Force as it confronts an uncertain future. This section also suggests how senior leaders might think about creating a force that is suited to address the challenges that these contexts will present. This is the unique contribution of the *Joint Operating Environment* to the broader discussion about the future. Before concluding, Part V offers some "leading questions" about topics that may fall outside the traditional purview of this study, but that nonetheless have important implications for the future Joint Force.

We will find ourselves caught off guard by changes in the political, economic, technological, strategic, and operational environments. We will find ourselves surprised by the creativity and capability of our adversaries. Our goal is not to eliminate surprise – that is impossible. Our goal is, by a careful consideration of the future, to suggest the attributes of a joint force capable of adjusting with minimum difficulty when the surprise inevitably comes. The true test of military effectiveness in the past has been in the ability of a force to diagnose the conditions it actually confronts and then quickly adapt. In the end, it will be our imagination and agility to envision and prepare for the future, and then to adapt to surprises, that will determine how the Joint Force will perform over the next twenty-five years. The agility to adapt to the reality of war, its political framework, and to the fact the enemy also consists of adaptive humans, has been the key component in military effectiveness in the past and will continue to be so in the future.

PART I: THE CONSTANTS

> *In the late fifth century BC, Athenian negotiators, speaking to their Spartan competitors, with whom they were soon at war, staked out their rationale for their refusal to abandon their position as Greece's other great power: "We have done nothing extraordinary, nothing contrary to human nature in accepting an empire when it was offered to us and then in refusing to give it up. Three very*

powerful motives prevent us from doing so – security, honour, and self interest. And we were not the first to act in this way. Far from it."[4]

Thucydides

A. The Nature of War

We cannot predict exactly what kind of war, or for what purposes, the armed forces of the United States will find themselves engaged in over the next quarter century. We can only speculate about possible enemies and the weapons they will bring to the fight. But we can state with certainty that the fundamental nature of war will not change. In a democracy such as the United States, political aims, pressures, and hesitations have always conditioned military operations – and will continue to do so. "When whole communities go to war... the reason always lies in some political situation."[5] War is a political act, begun for political purposes. In the twenty-first century war will retain its political dimension, even when it originates in the actions of non-state and transnational groups.

The Joint Force will operate in an international environment where struggle predominates. While the origins of war may rest on policy, a variety of factors has influenced the conduct of that struggle in the past and will do so in the future. The tension between rational political calculations of power on one hand and secular or religious ideologies on the other, combined with the impact of passion and chance, makes the trajectory of any conflict difficult if not impossible to predict. In coming decades, Americans must struggle to resist judging the world as if it operated along the same principles and values that drive our own country. In many parts of the world, there are no rational actors, at least in our terms. Against enemies capable of mobilizing large numbers of young men and women to slaughter civilian populations with machetes or to act as suicide bombers in open markets; enemies eager to die, for radical ideological, religious, or ethnic fervor; enemies who ignore national borders and remain unbound by the conventions of the developed world; there is little room for negotiations or compromise. It can become a matter of survival when human passion takes over. Such a world has existed in recent history – in World War II on the Eastern Front and on the islands of the Pacific, in Africa in the Rwandan genocide, and to some extent in Iraq. In a world where passions dominate, the execution of rational strategy becomes extraordinarily difficult.

War more than any other human activity engages our senses: at times providing a "rush" of fear, horror, confusion, rage, pain, helplessness, nauseous anticipation, and hyper-awareness. It is in these vagaries that imponderables and miscalculations accumulate to paralyze the minds of military and political leaders. In the cauldron of war, "It is the exceptional [human being] who keeps his powers of quick decision intact."[6]

There are other aspects of human conflict that will not change no matter what advances in technology or computing power may occur: fog and friction will distort, cloak, and twist the course of events. Fog will result from information overload, our own misperceptions and faulty assumptions, and the fact that the enemy will act in an unexpected fashion. Combined with the fog of war will be its frictions - that almost infinite number of seemingly insignificant incidents and actions that can go wrong, the impact of chance, and the horrific effect of combat on human perceptions. It will arise "from fundamental aspects of the human condition and unavoidable unpredictabilities that lie at the very core of combat processes."[7]

It is the constant fog and friction of war that turn the simple into the complex. In combat, people make mistakes. They forget the basics. They become disoriented, ignoring the vital to focus on the irrelevant. Occasionally, incompetence prevails. Mistaken assumptions distort situational awareness. Chance disrupts, distorts, and confuses the most careful of plans. Uncertainty and unpredictability dominate. Thoughtful military leaders have always recognized that reality, and no amount of computing power will eradicate this basic messiness.

Where friction prevails, tight tolerances, whether applied to plans, actions, or materiel are an invitation to failure – the more devastating for being unexpected. Operational or logistical concepts or plans that make no allowance for the inescapable uncertainties of war are suspect on their face – an open invitation to failure and at times defeat.

Still another enduring feature of conflict lies in the recurring fact that military leaders often fail to recognize their enemy as a learning, **adaptive force. War** "is not the action of a living force upon a lifeless mass... but always the collision of **two living forces.**"[8] Those living forces possess all the cunning and intractable characteristics human beings have enjoyed since the dawn of history.

Even where adversaries share a similar historical and cultural background, the mere fact of belligerence guarantees profound differences in attitudes, expectations, and behavioral norms. Where different cultures come into conflict, the likelihood that adversaries will act in mutually incomprehensible ways is even **more likely. Thus, "if you know the enemy and know yourself you need not fear the results of a hundred battles."**[9] The conduct of war demands a deep understanding of the enemy – his culture, history, geography, religious and ideological motivations, and particularly the huge differences in his perceptions of the external world. The fundamental nature of war will not change.

B. The Nature of Change

If war will remain a human endeavor, a conflict between two learning and adapting forces, changes in the political landscape, adaptations by the enemy, and advances in technology will change the character of war. Leaders are often late to recognize such changes. Driven by an inherent desire to bring order to a disorderly, chaotic universe, human beings tend to frame their thoughts about the future in terms of continuities and extrapolations from the present and occasionally the past. But a brief look at the past quarter century, to say nothing of the past four thousand years, suggests the extent of changes that coming decades will bring.

Twenty-five years ago the Cold War encompassed every aspect of the American military's thinking and preparation for conflict – from the strategic level to the tactical. Today, that all-consuming preoccupation is an historical relic. A quarter century ago, the United States confronted the Soviet Union, a truculent, intractable opponent with leaders firmly committed to the spread of Marxist-Leninist ideology and expansion of their influence. At that time, few in the intelligence communities or even among Sovietologists recognized the deepening internal crisis of confidence that would lead to the implosion of the Soviet Empire. The opposing sides had each deployed tens of thousands of nuclear weapons, as well as vast armies, air forces, and navies across the globe. Soviet forces were occupying Afghanistan and appeared on the brink of crushing an uprising of ill-equipped, ill-trained guerrillas. In El Salvador, a Soviet-backed insurgency was on the brink of victory.

Beyond the confrontation between the United States and Soviet Union lay a world that differed enormously from today. China was only emerging from the **dark years of Mao's rule. To China's south, India** remained mired in an almost medieval level of poverty, from which it appeared unlikely to escape. To the sub-continent's **west, the Middle East** was as plagued by political and religious troubles as today. But no one could have predicted then that within 25 years the United States would wage two major wars against Saddam Hussein's regime and commit much of its ground power to suppressing simultaneous insurgencies in Iraq and Afghanistan.

The differences between the culture and organization of the American military then and now further underline the extent of the disruptions with the past.

The lack of coordination among the forces involved in overthrowing the "New Jewel" movement in Grenada in October 1983 reminds us that at the time jointness was a concept honored more in the breach than observance. That situation led to the Goldwater-Nichols Act in 1986.

In terms of capabilities, stealth did not yet exist outside of the research and development communities. The M-1 Tank and the Bradley Fighting Vehicle were only starting to reach the Army's forward deployed units. The Global Positioning System (GPS) did not exist. The training ranges of the National Training Center, Twenty-Nine Palms, Fallon, and Nellis were just beginning to change U.S. preparations for war. Precision attack was largely a matter for tactical nuclear weapons.

STRATEGIC ESTIMATES IN THE TWENTIETH CENTURY

1900 If you had been a strategic analyst for the world's leading power, you would have been British, looking warily at Britain's age old enemy: France.

1910 You would now be allied with France, and the enemy would now be Germany

1920 Britain and its allies had won World War I, but now the British found themselves engaged in a naval race with its former allies the United States and Japan.

1930 For the British, naval limitation treaties were in place, the Great Depression had started and defense planning for the next five years assumed a "ten year" rule – no war in ten years. British planners posited the main threats to the Empire as the Soviet Union and Japan, while Germany and Italy were either friendly or no threat.

1936 A British planner would now posit three great threats: Italy, Japan, and the worst, are surgent Germany, while little help could be expected from the United States.

1940 The collapse of France in June left Britain alone in a seemingly hopeless war with Germany and Italy with a Japanese threat looming in the Pacific. America had only recently begun to scramble to rearm its military forces.

1950 The United States was now the world's greatest power, the atomic age had dawned, and a "police action" began in June in Korea that was to kill over 36,500 Americans, 58,000 South Koreans, nearly 3,000 Allied soldiers, 215,000 North Koreans, 400,000 Chinese, and 2,000,000

	Korean civilians before a cease-fire brought an end to the fighting in 1953. The main opponent in the conflict would be China, **America's ally in the war against Japan.**
1960	Politicians in the United States were focusing on a missile gap that did not exist; massive retaliation would soon give way to flexible response, while a small insurgency in South Vietnam hardly drew American attention.
1970	The United States was beginning to withdraw from Vietnam, its military forces in shambles. The Soviet Union had just crushed incipient rebellion in the Warsaw Pact. Détente between the Soviets and Americans had begun, while the Chinese were waiting in the wing to create an informal alliance with the United States.
1980	The Soviets had just invaded Afghanistan, while a theocratic revolution in Iran **had overthrown the Shah's regime.** "Desert One" -- an attempt to free American hostages in Iran -- ended in a humiliating failure, another indication of what pundits were calling "**the hollow force.**" America was the greatest creditor nation the world had ever seen.
1990	The Soviet Union collapses. The supposedly hollow force shreds the vaunted Iraqi Army in less than 100 hours. The United States had **become the world's greatest debtor nation.** No one outside of the Department of Defense has heard of the internet.
2000	Warsaw is the capital of a North Atlantic Treaty Organization (NATO) nation. **Terrorism is emerging as America's greatest threat.** Biotechnology, robotics, nanotechnology, HD energy, etc. are advancing so fast they are beyond forecasting
2010	Take the above and plan accordingly! What will be the disruptions of the next 25 years?

One might also note how much the economic and technological landscapes outside of the military have changed. Economically, in 1983 globalization was in its first stages and largely involved trade among the United States, Europe, and Japan. The tigers of Southeast Asia were emerging, but the rest of the world seemed caught in inescapable poverty. Just to give one example: in 1983 the daily transfer of capital among international markets was approximately $20 billion. Today, it is $1.6 trillion.

On the technological side, the internet existed only in the Department of Defense; its economic and communications possibilities and implications were not apparent. Cellular phones did not exist. Personal computers were beginning to come into widespread use, but their reliability was terrible. Microsoft was just emerging from Bill Gates' garage, while Google existed only in the wilder writings of science fiction writers. In other words, the revolution in information and communications technologies, taken for granted today, was largely unimaginable in 1983. A revolution had begun, but its implications remained uncertain and unclear. Other advances in science since 1983, such as the completion of the human genome project, nano technologies, and robotics, also seemed the provenance of writers of science fiction.

In thinking about the world's trajectory, we have reason to believe that the next twenty-five years will bring changes just as dramatic, drastic, and disruptive as those that have occurred in the past quarter century. Indeed, the pace of technological and scientific change is increasing. Changes will occur throughout the energy, financial, political, strategic, operational, and technological domains. While some change is foreseeable, even predictable, future joint force planning must account for the certainty that there will be surprises. How drastic, how disruptive they might be is at present not discernible and in some cases it will not even be noticed until they happen.

The interplay between continuities and disruptions will demand a joint force that can see both what has changed and what endures. The force must then have the ability to adapt to those changes while recognizing the value of fundamental principles. That can only result from an historically-minded mentality that can raise the right questions.

C. The Challenge of Disruptions

Trends may suggest possibilities and potential directions, but they are unreliable for understanding the future, because they interact with and are influenced by other factors. The down turn of Wall Street after the crash of 1929 might well have remained a recession, but passage of the Smoot-Hawley tariffs destroyed American trade with other nations and turned the recession into a catastrophic global depression. In considering the future, one should not underestimate the ability of a few individuals, even a single person, to determine the course of events. One may well predict that human beings will act in similar patterns of behavior in the future, but when, where and how remains entirely unpredictable. The rise of a future Stalin, Hitler, or Lenin is entirely possible, but completely unpredictable, and the context in which they might reach the top is unforeseeable.

> ### THE FRAGILITY OF HISTORY – AND THE FUTURE...
>
> The patterns and course of the past appear relatively straight-forward and obvious to those living in the present, but only because the paths not taken or the events that might have happened, did not. Nothing makes this clearer than the fates of three individuals in the first thirty plus years of the twentieth century. Adolf Hitler enlisted in the 16th Bavarian Reserve Regiment (the "List" Regiment) in early August 1914; two months later he and 35,000 ill-trained recruits were thrown against the veteran soldiers of the British Expeditionary Force. In one day of fighting the List Regiment lost one third of its men. When the Battle of Langemark was over, the Germans had suffered approximately 80% casualties. Hitler was unscratched. Seventeen years later, when Winston Churchill was visiting New York, he stepped off the curb without looking in the right direction and was seriously injured. Two years later in February 1933, Franklin Roosevelt was the target of an assassination attempt, but the bullet aimed for him, hit and killed the mayor of Chicago. Can any one doubt that, had any one of these three individuals been killed, the history of the twentieth century would have followed a fundamentally different course?

The interplay of economic trends, vastly different cultures and historical experiences, and the idiosyncrasies of leaders, among a host of other factors, provide such complexity in their interactions as to make prediction impossible. Winston Churchill caught those complexities best in his masterful history of World War I:

> *One rises from the study of the causes of the Great War with a prevailing sense of the defective control of individuals upon world fortunes. It has been well said, 'there is always more error than design in human affairs.' The limited minds of the ablest men, their disputed authority, the climate of opinion in which they dwell, their transient and partial contributions to the mighty problem, that problem itself so far beyond their compass, so vast in scale and detail, so changing in its aspects – **all this must surely be considered**...*[10]

Thus, individuals, their idiosyncrasies, genius, and incompetence, are major actors in these disruptions. Perhaps the worst president in American history, James Buchanan, was followed by the greatest, Abraham Lincoln. Individuals invariably remain the prisoners of their cultural and historical frame of reference, which makes the ability to understand, much less predict, the actions of other states and other leaders difficult. But we should not allow this to discourage us from gaining as deep an understanding as possible of the historical influences of potential political and military leaders at the strategic, operational, and tactical level.

Clearly, not all disruptions occur through the actions of individual leaders. Great events, involving the overthrow of regimes, the collapse of economic systems, natural disasters, and great conflicts within or among states have taken the flow of history and channeled it into new and unforeseen directions. Such singularities are truly unpredictable, except that we can be sure that they will happen again. They will twist the future into new and unexpected directions. Here, the only strategy that can mitigate the impact of surprise is a knowledge of the past, an understanding of the present, and a balanced force that is willing and able to adapt.

D. Grand Strategy

> *As in a building, which, however fair and beautiful the superstructure, is radically marred and imperfect if the foundations be insecure -- so if the strategy be wrong, the skill of the general on the battlefield, the valor of the soldier, the brilliancy of victory, however otherwise decisive, fail of their effect.*[11]
>
> *Mahan*

Future joint force commanders will not make grand strategy, but they must fully understand the ends it seeks to achieve. They will have a role to play in suggesting how the Joint Force might be used and the means necessary for the effective use of joint forces to protect the interests of the United States. Thus, their professional, nuanced advice as military leaders is essential to the casting of effective responses to strategic challenges.

In the twentieth century the relationship in the United States between political vision and military leaders responsible for the execution of policy proved crucial in winning two world wars and the Cold War. Yet the dialogue and discourses between those responsible for casting grand strategy and those responsible for conducting military operations has always involved tension, because their perspectives of the world inevitably differ. In the future, joint force commanders must understand the ends of strategy in order to recommend the forces required (the means) to achieve those ends. And policy makers must be clear as to the strengths, limitations, and potential costs of the employment of military forces. The relationship between ends and means drives the logic of joint operations.

Only clear and unfettered military advice from commanders to policy makers can provide the understanding required to employ the Joint Force effectively.

PART II: TRENDS INFLUENCING THE WORLD'S SECURITY

> *Engage people with what they expect; it is what they are able to discern and confirms their projection. It settles them into predictable patterns of response, occupying their minds while you wait for the extraordinary moment -- that which they cannot anticipate.[12]*
>
> Sun Tzu

Trend analysis is the most fragile element of forecasting. The world's future over the coming quarter of a century will be subject to enormous disruptions and surprises, natural as well as man-made. These disruptions, and many other contiguous forces, can easily change the trajectory of any single trend. The *Joint Operating Environment* recognizes that many, if not all, of the trends and trajectories of the future will be non-linear. But for the purpose of analysis, it has used a traditional approach to examine many of the trends and utilized conservative estimates. For instance, demographically, it has used estimates from sources such as the U.S. Census Bureau. Economically, the *Joint Operating Environment* assumes growth rates for developed countries of 2.5% and 4.5% for developing countries, including China. It is in this manner that this study considers the trends below. In the final analysis, the value of the trends lies not in accurately predicting them, but in intuiting how they might combine in different ways to form more enduring contexts for future operations. Trend analysis can also help in identifying some indicators or signposts that one can use to "check" the path that the world takes into the future and make adjustments as necessary. Nevertheless, the resource and strategic implications of even a conservative and linear rate of increase possess consequences that suggest a dark picture of the future.

A. Demographics

A good place to begin the discussion of trends is demographics, because what is happening demographically today, unless altered by some catastrophe, has predictable consequences for the populations of regions and states. Equally important, it possesses implications for future strategic postures and attitudes. In total, the world will add approximately 60 million people each year and reach a total of 8 billion by the 2030s. Ninety-five percent of that increase will occur in developing countries. **The more important point is that the world's troubles will occur not only in the areas of abject poverty, but also to an even greater extent in developing countries, where the combination of demographics and economy permits populations to grow, but makes meeting rising expectations difficult.** Here, the performance of the global economy will be key in either dampening down or inflaming ethnically or religiously-based violent movements.

The developed world confronts the opposite problem. During the next 25 years population growth in the developed world will likely slow or in some cases decline. In particular, Russia's **population is currently declining by 0.5% annually,** and given Russian health and welfare profiles, there is every prospect that decline will continue, barring a drastic shift in social attitudes or public policy. As a **recent Center for Strategic International Studies (CSIS) report suggested,** "Russia needs to cope with a rate of population decline that literally has no historical precedent in the absence of pandemic."[13] **To Russia's west, a similar, albeit less** disastrous situation exists. Over all, European nations stopped replacing their losses to deaths in 2007, and despite considerable efforts to reverse those trends, there is little likelihood their populations will significantly increase by the 2030s. This raises serious concerns about the sustainability of economic growth in that region. It also has serious implications for the willingness of European societies to bear the costs involved in lives and treasure that the use of military force inevitably carries with it.

Likewise, Japan's **population** will fall from 128 million to approximately 117 million in the 2030s, but unlike the case of Russia this will result not from any inadequacy of Japanese medical services, **which are among the world's best, but from the collapse of Japan's birth** rate. The Japanese are taking serious steps to address their demographic decline, a fact which explains their major research and development efforts in the field of robotics as well as their shift to a capital-intensive economy.

Over the next quarter century, China's **population** will grow by 170 million, but its population will age significantly because of strict enforcement of the government's **edict of one child per family.** An additional demographic factor, which may impact on Chinese behavior, is the choice of many families to satisfy that limitation with a male child. How the resulting imbalance between young males and females will play out **by the 2030s in China's external and internal** politics is impossible to predict, because there are few historical analogues. Nevertheless, there are some indications of an increasing predilection to violence among Chinese youth, while there have been exuberant displays of nationalistic feeling among the young in response to criticisms of China's behavior in Tibet.

By the 2030s the U.S. population will climb by more than 50 million to a total of approximately 355 million. This growth will result not only from births in current American families, but also from continued immigration, especially from Mexico and the Caribbean, which will lead to major increases in America's Hispanic population. By 2030 at least 15% of the population of every state will be Hispanic in origin, in some states reaching upwards of 50%. How effective Americans prove in assimilating these new immigrants into the nation's politics and culture will play a major role in America's prospects. In this regard, the historical ability of the United States to assimilate immigrants into its society and culture gives it a distinct advantage over most other nations, who display little willingness to incorporate immigrant populations into the mainstream of their societies.

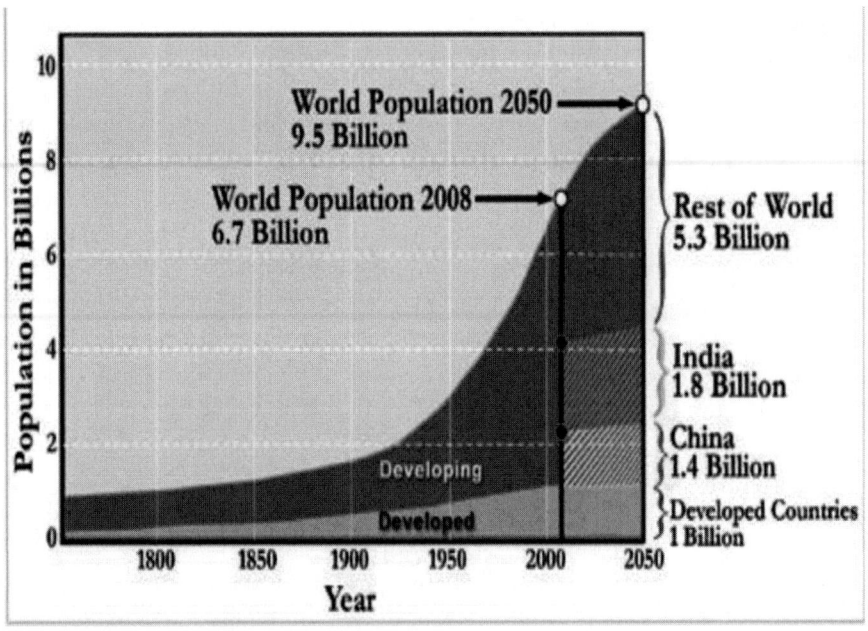

Source: United Nations Populations Reference Bureau

Population to 2050: Developed and Developing World

WORLD POPULATION PYRAMIDS

A population pyramid is a demographer's tool used to track the size and age composition of a country or group. Each bar represents an age group in four-year increments (youngest at the bottom) with males on the left and females on the right. The pyramids above show projected populations of selected countries in the 2030 time frameand the width of each pyramid is to scale. Thus, we see a 2030 Yemen that rivals Russia in terms of population. Developed countries generally show a typical "inverted" pattern with dramatic declines in the raw numbers of youth relative to the retired. This pattern of decline will be difficult to manage as most welfare systems in the developed world are based on an assumption of moderate population growth. Developing countries such as Nigeria and Yemen illustrate how the population pyramid in fact got its name, and are typical of fast-growing countries with large multi-children families. The effects of China's one-child policy are clear, especially when compared to fast-growing India. The United States occupies a middle position among states, with a large, yet relatively stable population.

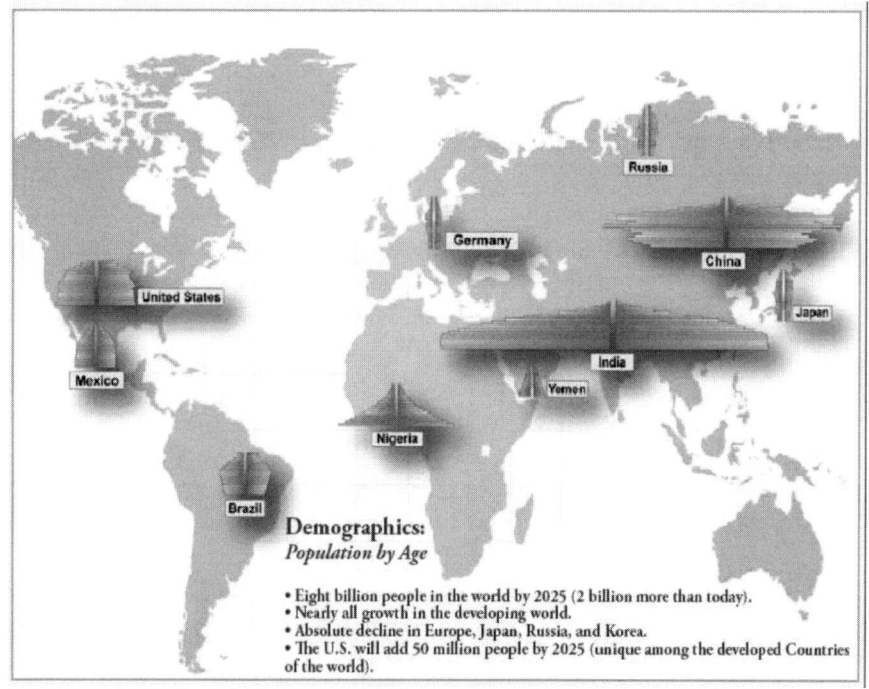

Demographics:
Population by Age

- Eight billion people in the world by 2025 (2 billion more than today).
- Nearly all growth in the developing world.
- Absolute decline in Europe, Japan, Russia, and Korea.
- The U.S. will add 50 million people by 2025 (unique among the developed Countries of the world).

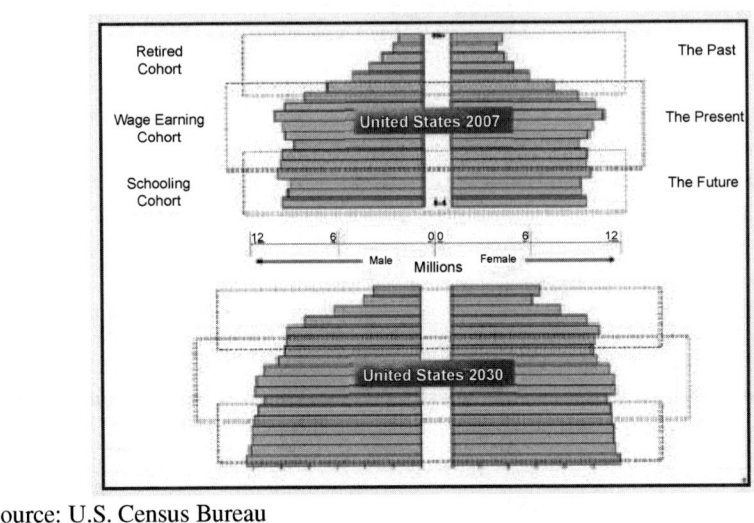

Source: U.S. Census Bureau

India will grow by 320 million during the next quarter of a century. The tensions that arise from a growing divide between rich and poor in a nation already driven by a multiplicity of races and religions could seriously impact on its potential for further economic growth. Exacerbating tensions will be the divide between the sub-continent's huge middle class and those in the villages mired in poverty, as well as the divide between Muslims and Hindus. Nevertheless, India's democratic system gives some latitude for political changes to accommodate society's poor.

The continued population growth Middle East and in Sub-Saharan Africa has only recently begun abating, but not fast enough to forestall a demographic crisis, where economic growth fails to keep pace with population growth. In areas of abject poverty, continued growth among the youth has significance for the employment of U.S. forces called upon to feed the starving and mitigate the suffering. Where economic growth fuels but does not satisfy expectations, the potential for revolution or war, including civil war, will be significant.

Even as the developing world copes with its youth bulge, the developed world will confront its acute aging problem. By the 2030s the number of elderly people in developed countries will double. In Japan there will be 63 elderly for every 100 workers, with Europe not far behind with 59 per 100. The United States will be slightly better off with 44 elderly per 100 workers. Even China will see its ratio of elderly to working population double (from 12 to 23 per 100 workers) as a result of better diet and improved medical care. Such demographic trends will make it less likely that nations in the developed world will sacrifice their youth in military adventures, unless extraordinary threats appear. Regions such as the Middle East and Sub-Saharan Africa, where the youth bulge will reach over 50% of the population, will possess fewer inhibitions about engaging in conflict.

Around the world, humanity is on the move, with Muslims and Africans moving to Europe, ethnic Chinese moving into Siberia, Mexicans and other Latin Americans moving north to the United States inequality of rich and poor. In some worst-case and Canada, and citizens of the Philippines and India scenarios, they portray the rise of resentment and providing the labor and small commercial backbones violence throughout the world as a direct result of the economies of the Gulf States. Equally important are the migrations occurring in war torn areas in Africa in areas like the Sudan, Somalia, Darfur, Rwanda. Such migrations disrupt patterns of culture, politics, and economics and in most cases carry with them the potential of further dislocations and troubles.

Everywhere, people are moving to cities. Skilled workers, doctors, and engineers are leaving the undeveloped world as fast as they can to make a living in the developed world. Increasingly, these global diasporas connect through the internet and telephone to their home countries. Often, the money they send back to their families forms major portions of the local economies back in their home communities.

B. Globalization

For the most part, the developed world recognizes that it has a major stake in the continuing progress of globalization. The same can be said for those moving into the developed world. Nevertheless, one should not ignore the histories and passions of popular opinion in these states as they make their appearance. One should not confuse developed world trappings for an underlying stability and maturity of civil societies. A more peaceful cooperative world is only possible if the pace of globalization continues. In particular, this means engaging China and other nations politically and culturally as they enter into the developed world.

The critics of globalization often portray its dark side in the inequality of rich and poor. In some worst-case scenarios, they portray the rise of resentment and violence throughout the world as a direct result of globalization. Not surprisingly, the future is likely to contain both good and bad as globalization accelerates the pace of human interaction and extends its reach.

LESSONS FROM THE HISTORY OF GLOBALIZATION

How can one best define globalization? Some might delineate it in terms of increased international trade, limited restrictions on the movement of peoples, and light regulation on the flow of capital. At least that was how politicians and pundits defined it at the start of the twentieth century. At that time, Europeans did not require passports to travel from one country to another on the continent, a situation restored only in the late 1990s. By 1913 the value of international trade as a percentage of world GDP had reached a level the global economy would not replicate until the last decade of the twentieth century. The economies of the United States and the German Reich were expanding at unheard of rates. Western merchants were queuing up to supply China's teeming masses, as that country opened its markets for the first time in centuries. Furthermore, the largest migration – and a peaceful one at that – in history was taking place, as 25 million Europeans left home, most immigrating to the United States.

The world also saw technological and scientific revolutions unequaled in history, which in turn spawned revolutions in travel and communications. Travel across the Atlantic was now a matter of days rather than weeks or months. Telegraph cables linked the continents for near instantaneous communications. Railroads allowed travelers to cross continents in days rather than months. The internal combustion engine was already impacting on travel by land, while the appearance of the aircraft in 1903 suggested even greater possibilities. A complex web of international agreements, such as the International Postal Union and the International Telegraph Conventions, welded these changes together. Again as with today, many were not content to leave the direction of the new world order to governments. In the first decade of the century activists formed 119 international organizations and 112 in the second decade.

For much of humanity, this was a time of hope and optimism. As early as the mid-nineteenth century, John Bright, a British industrialist, argued that "nothing could be so foolish as a policy of war for a trading nation. Any peace was better than the most successful war." In 1911 a British journalist, Norman Angell, published a work titled *The Great Illusion*, which became an international best seller. In it, he argued the expansion of global commerce had changed the nature of wealth, which no longer would depend on control of territory or resources.

> For Angell, the belief that military strength was the basis for security represented a dangerous illusion. As for war itself, it represented a futile endeavor incapable of creating material wealth, while putting much at risk. His arguments boiled down to a belief that the interlocking networks of global trade made war impossible. In 1913, he published an improved edition to even greater acclaim. Yet, within a year the First World War had broken out. The result of that conflict in political and economic terms was to smash globalization for the next seventy years. Angell had been right about the absolute destructive effects of modern war. He had been wrong about human nature and its passions.
>
> Why is this important? Because these same arguments have regained currency. For many, particularly in the West, the interlocking trading and communications networks of the twenty-first century with their benefits have made war, if not impossible, then at least obsolete. Accordingly, any future war would cost so much in lives and treasure that no rational political leader would ever pursue it. The problem is that rationality, at least in their terms, does not exist in much of the world outside Europe, the United States, and Japan. Saddam Hussein managed to invade two of Iraq's **six neighbors in the** space of less than ten years and sparked three wars in the period he ruled. The first of his wars against Iran resulted in approximately 250 thousand Iraqi deaths and half a million Iranian dead, while his wars against his own people killed upwards of 100 thousand. In historical terms, globalization is not the norm for human affairs.

The processes propelling globalization over the next two decades could **improve the lives of most of the world's population**, particularly for hundreds of millions of the poorest. Serious violence, resulting from economic trends, has almost invariably arisen where economic and political systems have failed to meet rising expectations. A failure of globalization would equate to a failure to meet those rising expectations. Thus, the real danger in a globalized world, where even the poorest have access to pictures and media portrayals of the developed world, lies in a reversal or halt to global prosperity. Such a possibility would lead individuals and nations to scramble for a greater share of shrinking wealth and resources, as occurred in the 1930s with the rise of Nazi Germany in Europe and Japan's "co-prosperity sphere" in Asia.

Admittedly, some will also be left behind by globalization, either through the misfortunes of geography, culture (much of sub-Saharan Africa), or design (North Korea and Burma). Many of these nations will be weak and failing states and will require an international array of economic, diplomatic, and military resources to establish or sustain stability.

In a globalized world of great nations, the United States may not always have to take the lead in handling the regional troubles that will arise. By the 2030s, every region of the world will likely contain local economic powers or regional organizations capable of leadership. In any case, the United States will often find it prudent to play a cooperative or supporting role in military operations around the world. In most cases the assisting of, or intervention in, failing states will require a cooperative engagement between the United States and regional powers. Again, the skills of a diplomat in working with other people and military organizations from different cultures must be in the tool kit of commanders, staffs, and personnel throughout the Joint Force.

C. Economics

Using a base line of 2.5% growth for the developed world and 4.5% growth for the developing world, including China and India (a figure that grossly understates the present growth trajectory of these two nations), the world economy would double by the 2030s from $35 trillion to $72 trillion. Global trade would triple to $27 trillion.

Given these projections, those living in extreme poverty would fall from 1.1 billion to 550 million, while those living on $2 a day would fall from 2.7 billion to 1.9 billion. Currently, only six countries in the developing world possess populations of over 100 million people and a GDP of at least $100 billion (China, Russia, India, Indonesia, Brazil, and Mexico). By the 2030s Bangladesh, Nigeria, Pakistan, the Philippines, and Vietnam will have joined that group. Thus, in terms of the developing world alone, there would be 11 states with the population and the economic strength to build military forces possessing the ability to project significant military power in their region.

As more young enter the work force, the developing world will need to increase employment by nearly 50 million jobs per year. China and India alone need to create 8 to 10 million jobs annually to keep pace with the numbers entering the work force every year. If economic growth suffices to provide such employment, it would go far to reduce international tensions and the endemic troubles inherent in youth bulges. While poverty has rarely been a driving force for revolutionary movements and wars, rising expectations often have. And in a world covered by media reports and movies from around the globe, rising expectations will increasingly be a driving force of politics, war, and peace, however well individual economies may perform.

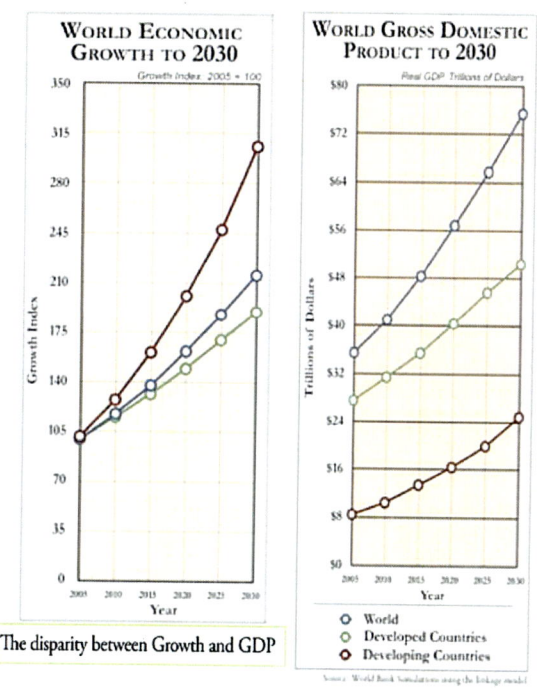

The disparity between Growth and GDP

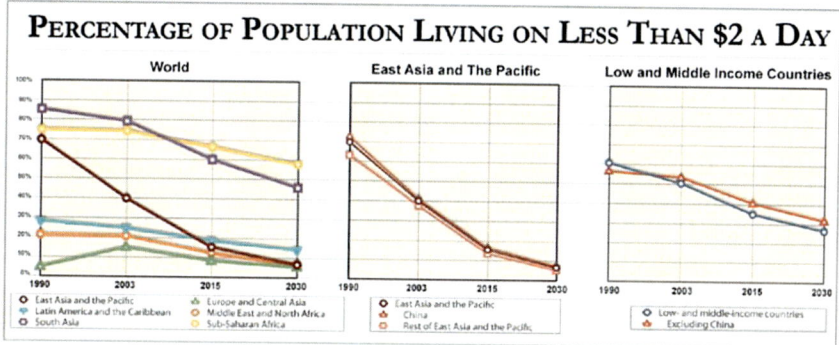

Source: World Bank

> ## THE VOLATILITY OF TRENDS
>
> Economic estimates rest on trend lines easily disputed both in the present and the future. For instance, were one to employ the same methodology used to compute the chart on the previous page in 1935, to predict future national GDPs in 1955, the results would be *off by an order of magnitude.* This chart presents the equivalent of a central scenario. Nevertheless, a word of warning is in order. In 1928 most economists would have given far rosier prospects for the American and world economies. Four years later, they would have given a far darker picture. That is the nature of change in economics as well as in very other human endeavor. Wide variations in either direction are not just feasible – they are likely.
>
> As the *Joint Operating Environment* goes to print the world is in the midst of the worst economic crisis since the Great Depression. While the final resolution is not yet in sight, the authors are of the opinion that the proactive measures taken by world governments (adding huge amounts of liquidity, recapitalizing the financial system and purchasing bad assets) will ensure that a global economic meltdown will not occur. Yet, it is almost certain that there will be a rather nasty global recession of indeterminate length. Recessions, while painful, are part of the natural business cycle and are unlikely to have a major impact on the trends outlined in this document.
>
> Nevertheless, the long-term strategic consequences of the current financial crises are likely to be significant. Over the next several years a new international financial order will likely arise that will redefine the rules and institutions that underpin the functioning, order, and stability of the global economy. There is one new watchword that will continue to define the global environment for the immediate future – interconnectedness.
>
> Until a new structure emerges, strategists will have to prepare to work in an environment where the global economic picture can change suddenly, and where even minor events can cause a cascading series of unforeseen consequences.

In contrast, real catastrophes may occur if economic growth slows or reverses either on a global scale or within an emerging power. Growing economies and economic hopes disguise a number of social ills and fractures. The results of a dramatic slowdown in China's **growth**, for example, are unpredictable and could easily lead to internal difficulties or aggressive behavior externally. That is precisely what happened in Japan in the early 1930s with the onset of the Great

Depression. Even within the most optimistic economic scenarios, there will be major areas of the world left behind – the bottom billion. Between now and the 2030s, many of these areas will likely lie in sub-Saharan Africa and the Middle East (excluding the oil boom countries). Although both regions have maintained impressive growth rates over the past several years, those rates have not been sufficient to decrease unemployment.

If economic stability and growth continue unabated up to the 2030s, there would be sufficient global resources to provide support for failing and failed states --that is, providing the political will is there. A broken economy is usually a harbinger of social collapse and anarchy, or ruthless despotism. Neither is attractive, but if the United States chooses to intervene in such situations, political and military leaders should keep in mind that they should only insert professional military forces if they are willing to sustain and inflict casualties which could result on both sides, as the experiences of the intervention in Somalia in 1993 underline.

A central component of America's global military posture is its massive economic power. This power is predicated on a financially-viable, globally connected domestic economy. Should this central feature of American power be weakened, it is highly likely that military capabilities will be diminished or otherwise degraded as a result.

D. Energy

To meet even the conservative growth rates posited above, global energy production would need to rise by 1.3% per year. By the 2030s, demand would be nearly 50% greater than today. To meet that demand, even assuming more effective conservation measures, the world would need to add roughly the equivalent of Saudi Arabia's current energy production every seven years.

Unless there is a major change in the relative reliance on alternative energy sources, which would require vast insertions of capital, dramatic changes in technology, and altered political attitudes toward nuclear energy, oil and coal will continue to drive the energy train. By the 2030s, oil requirements could go from 86 to 118 million barrels a day (MBD). Although the use of coal may decline in the Organization for Economic Cooperation and Development (OECD) countries, it will more than double in developing nations. Fossil fuels will still make up 80% of the energy mix in the 2030s, with oil and gas comprising upwards of 60%. The central problem for the coming decade will not be a lack of petroleum reserves, but rather a shortage of drilling platforms, engineers and refining capacity. Even were a concerted effort begun today to repair that shortage, it would be ten years before production could catch up with expected demand. The key determinant here would be the degree of commitment the United States and others would display in addressing the dangerous vulnerabilities the growing energy crisis presents.

That production bottleneck apart, the potential sources of future energy supplies nearly all present their own difficulties and vulnerabilities as shown here:

> ***Non-Organization of Petroleum Exporting Countries (OPEC) oil:*** New sources (Caspian Sea, Brazil, Colombia, and new portions of Alaska and the Continental shelf) could offset declining production in mature fields over the course of the next quarter century. But without drilling in currently excluded areas, they will add little additional capacity.
>
> ***Oil Sands and Shale:*** Production from these sources could increase from 1 MBD to over 4 MBD, but current legal constraints, such as U.S. law forbidding importation of oil from Canada's tar sands, discourage investment.
>
> ***Natural Gas:*** Production from this energy source could increase to the equivalent of 2 MBD, with half coming from OPEC countries.
>
> ***Biofuels:*** Production could increase to approximately 3 MBD–equivalent, but starting from a small base, biofuels are unlikely to contribute more than 1% of global energy requirements by the 2030s. Moreover, even that modest achievement could curtail the supply of foodstuffs to the world's growing population, which would add other national security challenge to an already full menu.
>
> ***Renewable:*** Wind and solar combined are unlikely to account for more than 1% of global energy by 2030. That assumes the energy from such sources will more than triple, which alone would require major investments.
>
> ***Nuclear:*** Nuclear energy offers one of the more promising technological possibilities, given significant advances in safety since the 1970s. In particular, it could play a major role in replacing coal–fired plants, and a greater supply of cheap electricity could encourage electric–powered transportation. Nevertheless, expanding nuclear plants confronts considerable opposition because of public fears, while the disposal of nuclear waste remains a political

hot potato. Moreover, construction of nuclear power plants in substantial numbers will take decades.

OPEC: To meet climbing global requirements, OPEC will have to increase its output from 30 MBD to at least 50 MBD. Significantly, no OPEC nation, except perhaps Saudi Arabia, is investing sufficient sums in new technologies and recovery methods to achieve such growth. Some, like Venezuela and Russia, are actually exhausting their fields to cash in on the bonanza created by rapidly rising oil prices.

None of the above provides much reason for optimism. At present, the United States possesses approximately 250 million cars, while China with its immensely larger population possesses only 40 million.

The Chinese are laying down approximately 1,000 kilometers of four–lane highway every year, a figure suggesting how many more vehicles they expect to possess, with the concomitant rise in their demand for oil. The presence of Chinese "civilians" in the Sudan to guard oil pipelines underlines China's concern for protecting its oil supplies and could preview a future in which other states intervene in Africa to protect scarce resources.

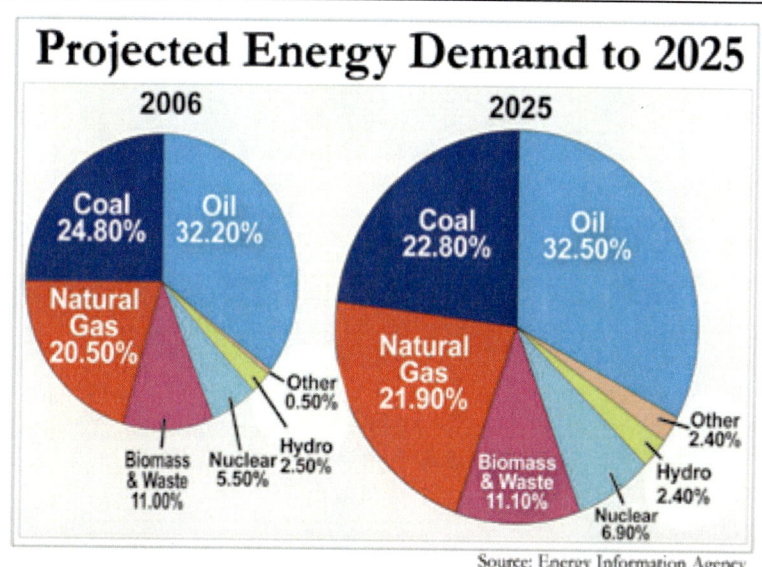

Source: Energy Information Agency

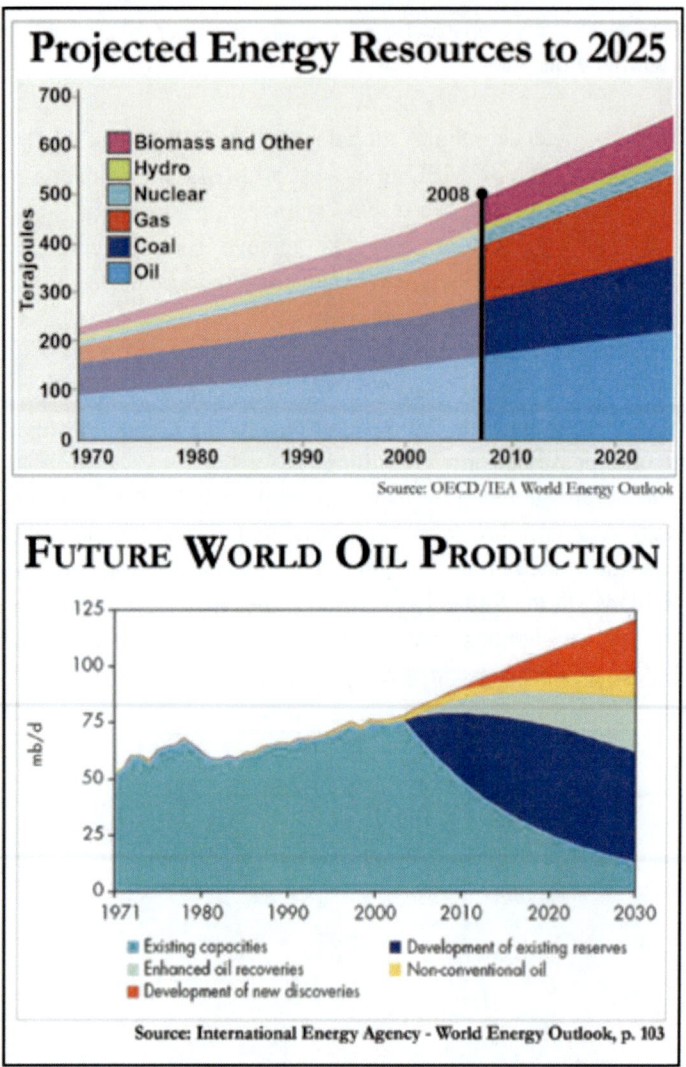

Although the world depends on oil, existing capacities and the development of existing reserves cannot keep up with demand. Massive investments in enhanced oil recovery techniques, nonconventional oil reserves such as oil shale, and large scale new finds will be required to meet anticipated future oil demand.

> **In summary:**
>
> To generate the energy required worldwide by the 2030s would require us to find an additional 1.4 MBD every year until then.
>
> During the next twenty-five years, coal, oil, and natural gas will remain indispensable to meet energy requirements. The discovery rate for new petroleum and gas fields over the past two decades (with the possible exception of Brazil) provides little reason for optimism future efforts will find major new fields.
>
> At present, investment in oil production is only beginning to pick up, with the result that production could reach a prolonged plateau. By 2030, the world will require production of 118 MBD, but energy producers may only be producing 100 MBD unless there are major changes in current investment and drilling capacity.
>
> ***By 2012, surplus oil production capacity could entirely disappear, and as early as 2015, the shortfall in output could reach nearly 10 MBD.***
>
> To avoid a disastrous energy crunch, together with the economic consequences that would make even modest growth unlikely, the developed world needs to invest heavily in oil production. There appears to be little propensity to consider such investments. Although as oil prices increase, market forces will inexorably create incentives. But the present lack of investment in this area has created major shortages in infrastructure (oil rigs, drilling platforms, etc.) necessary for major increases in exploration and production.

The implications for future conflict are ominous. If the major developed and developing states do not undertake a massive expansion of production and refining capabilities, a severe energy crunch is inevitable. While it is difficult to predict precisely what economic, political, and strategic effects such a shortfall might produce, it surely would reduce the prospects for growth in both the developing and developed worlds. Such an economic slowdown would exacerbate other unresolved tensions, push fragile and failing states further down the path toward collapse, and perhaps have serious economic impact on both China and India. At best, it would lead to periods of harsh economic adjustment. To what extent conservation measures, investments in alternative energy production, and efforts to expand petroleum production from tar sands and shale would mitigate such a period of adjustment is difficult to predict. One should not forget that the Great Depression spawned a number of ferocious totalitarian regimes that sought economic prosperity for their nations by ruthless conquest, while Japan went to war in 1941 to secure its energy supplies.

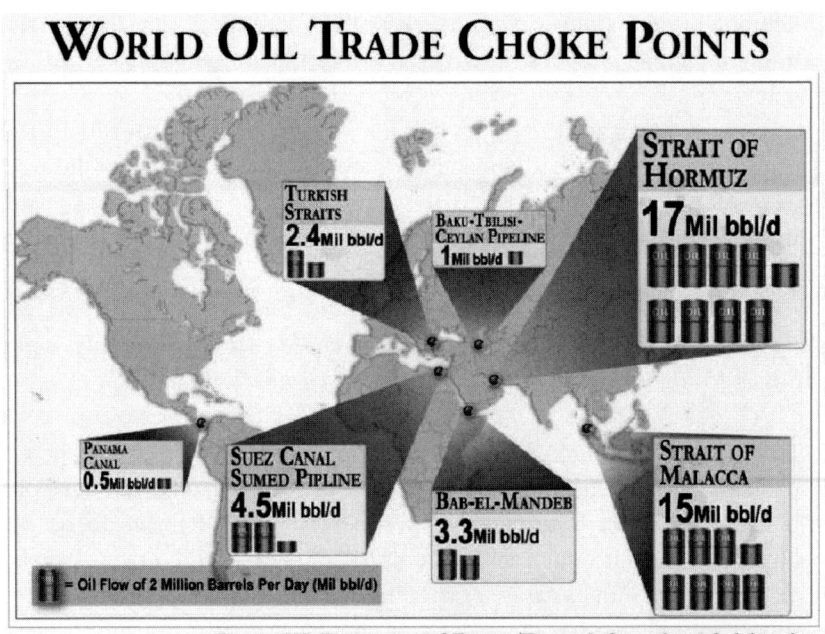

Source: U.S. Department of Energy/Energy Information Administration

OPEC nations will remain a focal point of great-power interest. These nations may have a vested interest in stymieing production increases, both to conserve finite supplies and keep prices high. Should one of the consumer nations choose to intervene forcefully, the "arc of instability" running from North Africa though to Southeast Asia easily could become an "arc of chaos," involving the military forces of several nations.

OPEC nations will find it difficult to invest much of the cash inflows that steadily rising oil prices bring. While they will invest substantial portions of such assets globally through sovereign wealth funds – investments that come with their own political and strategic difficulties – past track records, coupled with their appraisal of their own military weaknesses, suggests the possibility of a military buildup. With the cost of precision weapons expected to decrease and their availability increasing, joint force commanders could find themselves operating in environments where even small, energy-rich opponents have military forces with advanced technological capabilities. These could include advanced cyber, robotic, and even anti-space based systems.

Finally, presuming the forces propelling radical Islam at present do not dissipate, a portion of OPEC's windfall might well find its way into terrorist coffers, or into the hands of movements with deeply anti-modern, anti-Western goals, movements which have at their disposal increasing numbers of unemployed young men eager to attack their perceived enemies.

One other potential effect of an energy crunch could be a prolonged U.S. recession which could lead to deep cuts in defense spending (as happened during the Great Depression). Joint force commanders could then find their capabilities diminished at the moment they may have to undertake increasingly dangerous missions. Should that happen, adaptability would require more than preparations to fight the enemies of the United States, but also the willingness to recognize and acknowledge the limitations of America's military forces. The pooling of U.S. resources and capabilities with allies would then become even more critical. Coalition operations would become essential to protecting national interests.

E. Food

Two major factors drive food requirements: a growing global population and prosperity that expands dietary preferences. While food shortages still occur today, they are more likely to reflect politically-inflicted, rather than natural causes. Several mitigating trends could diminish the possibility of major food shortages.

For starters, any slowdown in the world's population growth may reduce overall demand for food and thus ease pressure to expand and intensify agriculture. On the other hand, increased animal protein use in countries with rapidly rising income levels is placing considerable pressure on the world's food supply, since animal production requires much greater input for calories produced. Opposition to genetically modified foods is dissipating. As a result, there is a reasonable chance of sparking a new "green revolution" that would expand crop and protein production sufficiently to meet world requirements. The main pressures on sufficient food supplies will remain in countries with persistently high population growth and a lack of arable land, in most cases exacerbated by desertification and shortages in rainfall.

In a world with adequate global supply but localized food shortages, the real problems will stem from how food is distributed. How quickly the world reacts to temporary food shortages inflicted by natural disasters will also pose challenges. In such cases, joint forces may find themselves involved in providing lift, logistics, and occasionally security to those charged with relief operations.

Natural disease will also have a say in the world's food supply. The Irish potato blight was not an exceptional historical event. As recently as 1954, 40% of America's wheat crop failed as a result of black-stem disease. There are reports of a new aggressive strain of this disease (Ug99) spreading across Africa and possibly reaching Pakistan. Blights threatening basic food crops such as potatoes and corn could have destabilizing effects on nations close to the subsistence level. Food crises have led in the past to famine, internal and external conflicts, the collapse of governing authority, migrations, societal collapse, and social disorder. In such cases, many people in the crisis zone may be well-armed and dangerous, making the task of the Joint Force in providing relief that much more difficult. In a society confronted with starvation, food becomes a weapon every bit as important as ammunition.

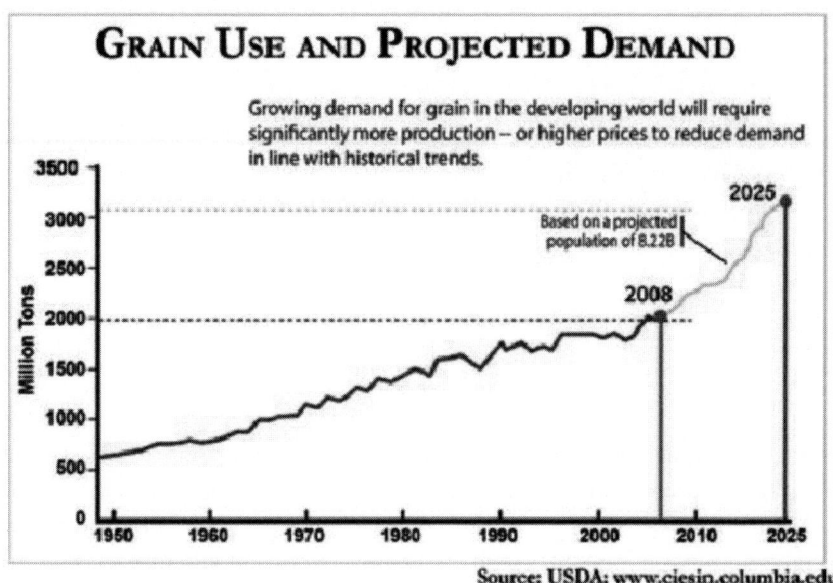

Access to fish stocks has been an important natural resource for the prosperity of nations with significant fishing fleets. Competition over access to these resources has often resulted in naval conflict. Conflicts have erupted as recently as the Cod War (1975) between Britain and Iceland and the Turbot War (1995) between Canada and Spain. In 1996, Japan and Korea engaged in a naval standoff over rocky outcroppings that would establish extended fishing rights in the Sea of Japan. These conflicts saw the use of warships and coastal protection vessels to ram and board vessels, and open conflict between the naval forces of these states. Over-fishing and depletion of fisheries and competition over those that remain have the potential for causing serious confrontations in the future.

F. Water

As we approach the 2030s, agriculture will likely remain the source of greatest demand for water worldwide, accounting for 70% of total water usage. In comparison, industry will account for only 20%, while domestic usage will likely remain steady at 10%. Per unit harvest yield, developed nations are more efficient than developing nations in using available water supplies for agricultural irrigation and use far less than the 70% average. Improved agricultural techniques could further increase the amount of land under irrigation, and increase yields per unit of water used.

In short, from a global perspective, there should be more than sufficient water **to support the world's population** during the next quarter century. However, in some regions the story is quite different. The Near East and North Africa use far more than the global average of 70% of available water dedicated to irrigation. By the 2030s, at least 30 developing nations could use even more of their water for irrigation.

In recent times, the increasing unreliability of an assured supply of rain water has forced farmer stoturn more to groundwater in many areas. As a result, aquifer levels are declining at rates between one and three meters per year. The impact of such declines on agricultural production could be profound especially since aquifers, once drained, may not refill for centuries.

> *Within a quarter century, water scarcity could affect approximately 3 billion people.*

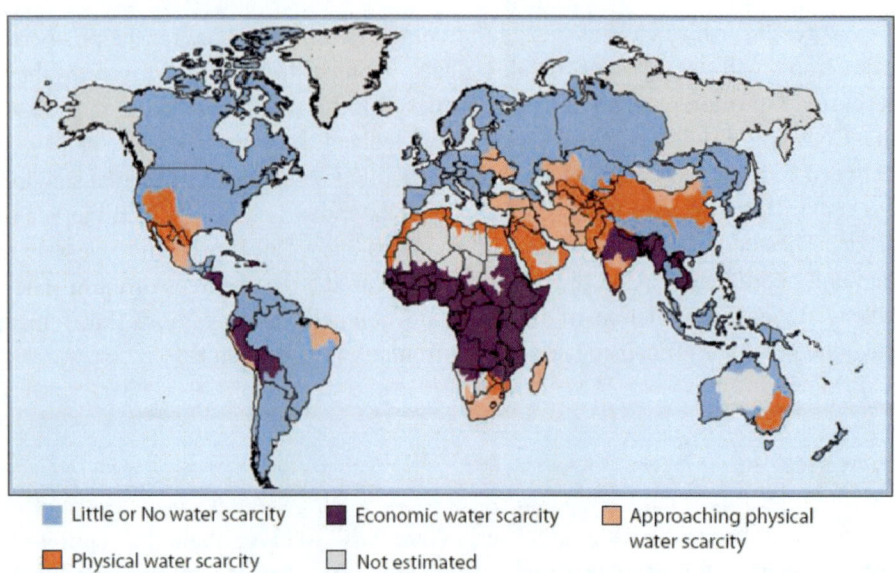

- Little or No water scarcity
- Physical water scarcity
- Economic water scarcity
- Not estimated
- Approaching physical water scarcity

Physical Scarcity: Physical access to water is limited, or sources of water overused and overmanaged, leading to serious water scarcity downstream.

Economic Scarcity: A population does not have adequate financial or political means to obtain adequate sources of water.

Source: International Water Management Institute

Projected Water Scarcity in 2025

One should not minimize the prospect of wars over water. In 1967, Jordanian and Syrian efforts to dam the Jordan River was a contributing cause of the Six-Day War between Israel and its neighbors. Today, Turkish dams on the upper Euphrates and Tigris Rivers, the source of water for the Mesopotamian basin, pose similar problems for Syria and Iraq. Turkish diversion of water to irrigate mountain valleys in eastern Turkey already reduces water downstream. Even though localized, conflicts sparked by water scarcity easily could destabilize whole regions. The continuing crisis in Sudan's **Darfur region**, now spreading to Chad, is an example of what could happen on a wider scale between now and the 2030s. Indeed, it is precisely along other potential conflict fault lines that potential crises involving water scarcity are most likely.

Whether the United States would find itself drawn into such conflicts is uncertain, but what is certain is that future joint force commanders will find conflict over water endemic to their world, whether as the spark or the underlying cause of conflicts among various racial, tribal, or political groups. Were they called on to intervene in a catastrophic water crisis, they might well confront chaos, with collapsing or impotent social networks and governmental services. Anarchy could prevail, with armed groups controlling or warring over remaining water, while the specter of disease resulting from unsanitary conditions would hover in the background.

The latter is only one potential manifestation of a larger problem. Beyond the problems of water scarcity, will be those associated with water pollution, whether from uncontrolled industrialization, as in China, or from the human sewage expelled by the mega-cities and slums of the world. The dumping of vast amounts of waste into the world's rivers and oceans threatens the health and welfare of large portions of the human race, to say nothing of the affected ecosystems. While joint forces rarely will have to address pollution problems directly, any operations in polluted urban areas will carry considerable risk of disease. Indeed, it is precisely in such areas that new and deadly pathogens are most likely to arise. Hence, commanders may be unable to avoid dealing with the consequences of chronic water pollution.

G. Climate Change and Natural Disasters

The impact of global warming and its potential to cause natural disasters and other harmful phenomena such as rising sea levels has become a prominent—and controversial—national and international concern. Some argue that there will be more and greater storms and natural disasters, others that there will be fewer.[14] In many respects, scientific conclusions about the causes and potential effects of global warming are contradictory.

Whatever their provenance, tsunamis, typhoons, hurricanes, tornadoes, earthquakes and other natural catastrophes have been and will continue to be a concern of joint force commanders. In particular, where natural disasters collide with growing urban sprawl, widespread human misery could be the final straw that breaks the back of a weak state. In the 2030s as in the past, the ability of U.S. military forces to relieve the victims of natural disasters could help the United States' image around the world. For example, the contribution of U.S. and partner forces to relieving the distress caused by the catastrophic Pacific tsunami of December 2006 reversed the perceptions of America held by many Indonesians. Perhaps no other mission performed by the Joint Force provides so much benefit to the interests of the United States at so little cost.

H. Pandemics

One of the fears haunting the public is the appearance of a pathogen, either man–made or natural, **able to devastate mankind, as the** "Black Death" **did in the Middle East and Europe in the middle of the fourteenth century. Within barely a year, approximately one–third of Europe's population** died. The second- and third-order effects of the pandemic on society, religion, and economics were devastating. In effect, the Black Death destroyed the sureties undergirding Medieval European civilization.

It is less likely that a pandemic on this scale will devastate mankind over the next two decades. Even though populations today are much larger and more concentrated, increasing the opportunities for a new pathogen to spread, the fact that mankind lives in a richer world with greater knowledge of the world of microbes, the ability to enact quarantines, a rapid response capability, and medical treatment, suggests that authorities could control even the most dangerous of pathogens. The crucial element in any response to a pandemic may be the political will to impose a quarantine.

The rapid termination of 2003's Severe Acute Respiratory Syndrome (SARS) pandemic does provide hope that current medical capabilities could handle most pandemic threats successfully. In the five months after initial reports from East Asia in February of an atypical respiratory disease, medical authorities reported more than 8,000 cases in 30 different countries. The disease itself was highly contagious and life-threatening: almost 10% of reported cases died. However, once doctors identified the disease, the combined efforts of local, national, and international authorities contained it. Newly reported cases increased rapidly in March and April 2003, peaked in early May, and thereafter rapidly declined.

The SARS case suggests that the risk of a global pandemic is not as great as some fear. That does not mean that it is nonexistent. A repetition of the 1918 influenza pandemic, which led to the deaths of millions world-wide would have the most serious consequences for the United States and the world politically as well as socially. The dangers posed by the natural emergence of a disease capable of launching a global pandemic are serious enough, but the possibility also exists that a terrorist organization might acquire a dangerous pathogen.

The deliberate release of a deadly pathogen, especially one genetically engineered to increase its lethality or virulence, would present greater challenges than a naturally occurring disease like SARS. While the latter is likely to have a single point of origin, terrorists would seek to release the pathogen at several different locations and it would spread faster. This would seriously complicate both the medical challenge of bringing the disease under control and the security task of fixing responsibility for its appearance.

The implications for joint forces of a pandemic as widespread and dangerous as that of 1918 would be profound. American and global medical capabilities would soon find themselves overwhelmed. If the outbreak spread to the United States, the Joint Force might have to conduct relief operations beyond assisting in law-enforcement and maintaining order when legal prerequisites are met, as currently identified by the National Strategy for Pandemic Influenza. Even as joint force commanders confronted an array of missions, they would have to take severe measures to preserve the health of their forces and protect medical personnel and facilities from public panic and dislocations. Thucydides captured the moral, political, and psychological dangers that a global pandemic would cause in his description of the plague's impact on Athens: "For the catastrophe was so overwhelming that men, not knowing what would happen next to them, became indifferent to every rule of religion or of law."[15]

I. Cyber

Perhaps the most important trend in the area of science and technology is the continuing information and communications revolution and its implications. Although many pundits have touted the ability of information to "lift the fog and friction of war," such claims have foundered on the rocks of reality.

- An iPod today can hold some 160 gigabytes of data, or 160,000 books. The iPod of 2020 could potentially hold some 16 terabytes of information – essentially the entire Library of Congress.
- Connectivity to the home (or node in military networks) grows by 50% a year. Therefore by 2030, people will have about 100,000 times more bandwidth than today.
- The computing capacity available to the average home will be a computer that runs at a rate of one million times faster than a computer today (2.5 petabytes vs. 2.5 gigabytes). A typical home computer would be capable of downloading the entirety of today's Library of Congress (16 terabytes), in 128 seconds – just over two minutes' time. The technical capacity of the telegraph in 1900, was some 2 bits per second across continental distances, meaning that same Library of Congress would have required a transmission time of 3,900 years.

Key to understanding information technology in the 2030s is the fact that the pace of technological change is accelerating almost exponentially. Because most individuals tend to view change in a linear fashion, they tend to overestimate what is achievable by technology in the short term, while dramatically underestimating and discounting the power of scientific and technological advances in the long term.

If the pace of technical advances holds true, greater technological change will occur over the next twenty years than occurred in the whole of the twentieth century. In many ways the world of 2030 will be nearly as strange as the world of 2000 would have been to an observer from 1900. The advances in communication and information technologies will significantly advance the capabilities of the Joint Force. Nevertheless, those same advances will be available to America's opponents and they will use those advances to attack, degrade, and disrupt communications and the flow of information. Indeed, our adversaries have often taken advantage of computer networks and the power of information technology not only to directly influence the perceptions and will of the United States, its decision-makers, and population, but also to plan and execute savage acts of terrorism. It is also essential that joint forces be capable of functioning in an information-hostile environment, so as not to create an Achilles' heel by becoming too network dependent.

EXPONENTIAL GROWTH OF COMPUTING

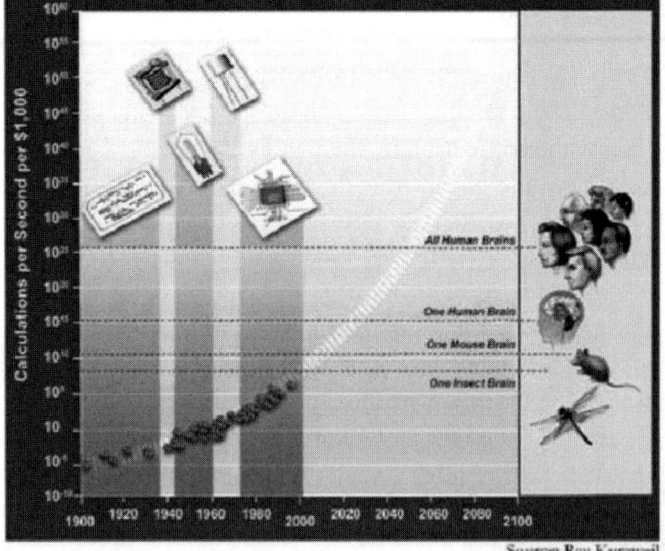

Source: Ray Kurzweil

J. Space

In 2007 the Chinese used an interceptor missile to destroy a satellite in space. In that single act, they made clear their belief that space was a potential theater of conflict and that they aimed to possess the capability to fight in that environment. As with the profusion of inexpensive precision weapons, technological advances and increasing wealth will place the ability to conduct military operations in space within the reach of an increasing number of players.

Over the past several decades the United States has enjoyed an unchallenged dominance over the dark realm beyond the atmosphere. However, the increasing proliferation of launch and satellite capabilities, as well as the development of anti-satellite capabilities, has begun to level the playing field. Other countries are leveraging the benefits of space for both commercial and defense applications, and the United States already confronts increased competition over its use. This will increasingly be the case over coming decades. The implications are clear: the Joint Force is going to have to be in a position to defend the spaced-based systems on which so many of its capabilities depend. As well, the Joint Force must anticipate the inevitable attack and know how to operate effectively when these attacks degrade those systems.

K. Conclusion

The previous discussion outlined just some of the trends that are likely to influence the security environment for the next quarter century. These individual trends, whether they adhere to predictions or not, will combine together in ways to form more broad and robust contexts that will define the world in which the Joint Force will operate in the future. By understanding the trends and resultant contexts, joint force leaders have a way to appreciate their implications, and to identify some key indicators to watch along the way. This provides a means of assessing our assumptions and predictions, and our progress towards building and operating the Joint Force to meet the future. What follows then is a discussion of the contextual world of the 2030s.

PART III: THE CONTEXTUAL WORLD

> *Contexts of conflict and war are the environment created by the confluence of major trends. Contexts illuminate why wars occur and how they might be waged.*[16]
>
> Colin Gray

A. Competition and Cooperation among Conventional Powers

Competition and conflict among conventional powers will continue to be the primary strategic and operational context for the Joint Force over the next 25 years. For the purpose of the *Joint Operating Environment*, a "conventional power" is an organization that is governed by, and behaves according to, recognized norms and codes – conventions. Such institutions may be political (the state), financial, judiciary, business and economic, academic, and many more. Conventions may include the Geneva Convention, the Law of Armed Conflict, United Nations Resolutions, National and International Law, international trade agreements, diplomatic alliances, monetary and banking conventions, and many more. These are groups that are broadly recognized as being legitimate actors, behaving according to broadly recognized rule sets.

The state will continue to be among the most powerful conventional institutions. It has become popular to suggest that the era of states is coming to an end. In fact states, in one form or another, have been the order of most of human affairs since the dawn of history in almost all cultures. The chaos in places such as Somalia, Sierra Leone, Afghanistan, and Iraq, where the state's structure has been dysfunctional for periods of time, is further testimony to the utility of a working state.

This is not to say that states will not vary from culture to culture, region to region. As well, the state will undoubtedly change in response to the influences of politics, geography, migration, economics and other factors. But though it will mutate and adapt to the international environment's changing conditions, the state will continue to survive as a centralized mechanism through which power is organized and which provides the internal and external security required by its citizens. Some aspects of globalization, and the related rise of non-state powers, will pose difficulties to states in their efforts to preserve their status, but the state will endure as a major power broker into the 2030s.

In the next 25 years, the relative balance of power between states will shift, some growing faster than the United States and many states weakening relative to the United States. The variables that affect the growth of states range from wars, to the effectiveness of political leaders, economic realities, ideological preconceptions, and ethnic and religious forces. All will to one extent or another influence the course of future events. Recognizing that reality, present trends suggest that the era of the United States as the sole superpower may be coming to an end. The charts on page 25, highlighting potential growth in various nations between 2008 and the 2030s, suggest much about the nature of the emerging international arena.

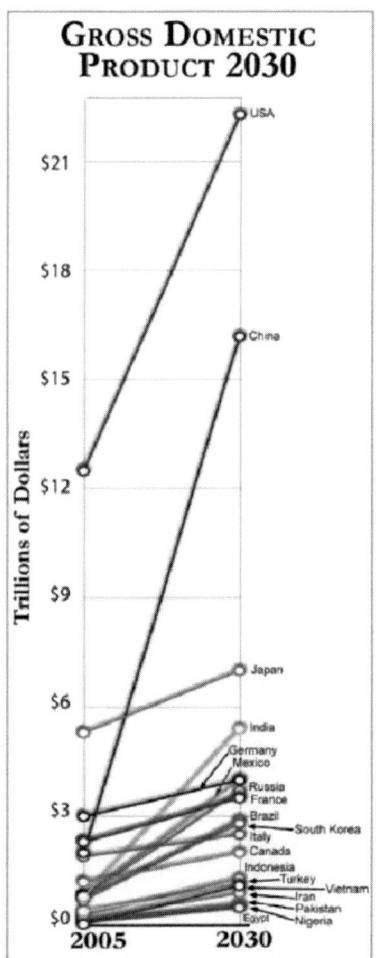
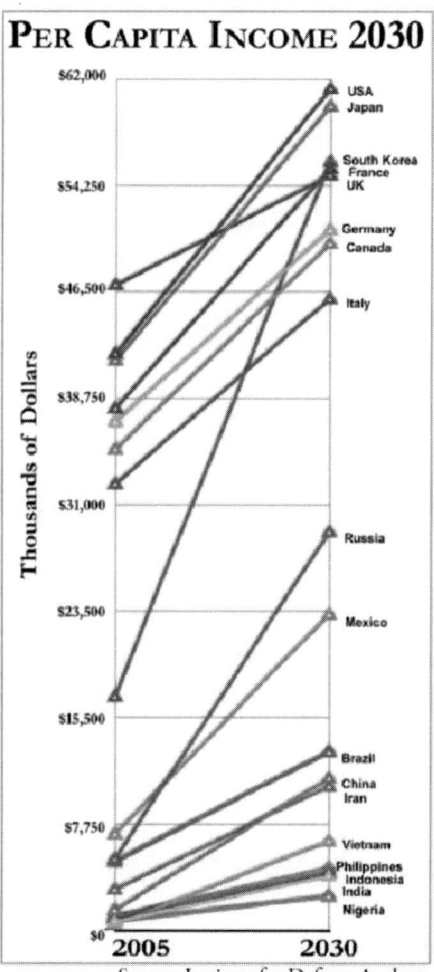

Source: Institute for Defense Analyses

While China's rise represents the most significant single event on the international horizon since the collapse of the Cold War, it is not the only story. Steady, if not rapid economic growth appears to be the norm for much of the world over the coming decades, provided sufficient energy remains available to fuel that growth. Russia and India are both likely to become richer, although Russia's strength is fragile, resting as it does on unfavorable demographic trends, a single-commodity (oil) economy, and a lack of serious investment in repairing its crumbling infrastructure.

As the figure satright suggest, based on a GDP per capita basis, a number of countries will be able to field larger conventional militaries over the course of coming decades.

Indeed, the story around the globe is one of substantial potential rearmament. While the rise of Nigeria, Turkey, Brazil, Vietnam, and Egypt may not be as dramatic as what is happening in South and East Asia, their increasing power is and will be remarkable. Admittedly these nations will likely not be able to field globally deployable forces, but they are in a position to build military forces which could either stabilize or destabilize their regions and could significantly challenge the ability of the United States to project military force into their area.

The critical issue will lie in national will. What has mattered most in the past has been the intent and national goals of states. In the 1930s, the democracies of Western Europe and the United States possessed the economic strength to crush Hitler's Germany, but lacked the will to rearm – they refused to see the threat. Today, many of these same countries make up the European Union and could field forces as large and capable as those of the United States, but again they lack the will. Since the end of the Cold War, many European nations have engaged in what could be classified as disarmament. The great question confronting Europeans is whether this trend will continue, or whether some impending threat – an aggressive and expansionist Russia, the internal stress of immigration, or radical Islamic extremism – will awaken them.

It is also conceivable that combinations of regional powers with sophisticated regional capabilities could band together to form a powerful anti-American alliance. It is not hard to imagine an alliance of small,cash-rich countries arming themselves with high-performance long-range precision weapons. Such a group could not only deny U.S. forces access into their country, but could also prevent American access to the global commons at significant ranges from their borders.

Not all conventional organizations will be states. Many transnational organizations will also behave according to a recognized set of conventional rules. Samuel Huntington describes the activity of these groups in this way:

Transnational organizations try to ignore [sovereignty]. While national representatives and delegations engage in endless debate at U.N. conferences and councils, the agents of transnational organizations are busily deployed across the continents, spinning the webs that link the world together.[17]

In this environment, the U.S. must strive to use its tremendous powers of inspiration, not just its powers of intimidation.[18] How America operates in this world of states and other conventional powers will be a major factor in its ability to project its influence and soft power beyond the long shadow cast by its raw military power. It will remain first among equals due to its military, political and economic strengths. But in most endeavors it will need partners, whether from traditional alliances or coalitions of the willing. Thus, the United States will need to sharpen its narrative about the unique vision we offer to the world and to inspire like-minded partners to strive and sacrifice for common interests. Alliances, partnerships, and coalitions will determine the framework in which joint force commanders operate. This will require diplomacy, cultural and political understanding, as well as military competencies. Here, the example that Dwight Eisenhower displayed as overall commander of Allied Forces that invaded Europe is particularly useful for future U.S. military leaders.

B. Potential Challenges and Threats

1. China

The Sino-American relationship represents one of the great strategic question marks of the next twenty-five years. Regardless of the outcome – cooperative or coercive, or both – China will become increasingly important in the considerations and strategic perceptions of joint force commanders.

The course that China takes will determine much about the character and nature of the twenty-first century -whether it will be "another bloody century,"[19] or one of peaceful cooperation. The Chinese themselves are uncertain as to where their strategic path to the future will lead. Deng Xiaoping's **advice for China to "disguise its ambition and hide its claws"** may represent as forthright a statement as the Chinese can provide. What does appear relatively clear is that the Chinese are thinking in the long term regarding their strategic course. Rather than emphasize the future strictly in military terms, they seem willing to see how their economic and political relations with the United States develop, while calculating that eventually, their growing strength will allow them to dominate Asia and the Western Pacific.

Source: Department of Defense

History provides some hints about the challenges the Chinese confront in adapting to a world where they are on a trajectory to become a great power. For millennia, China has held a position of cultural and political dominance over the lands and people on its frontiers that has been true of no other civilization. The continuities of Chinese civilization reach back to a time when the earliest civilizations in the Nile and the Mesopotamian valleys were emerging. But those continuities and the cultural power of China's civilization have also provided a negative side: to a considerable extent they have isolated China from currents and developments in the external world. China's history for much of the twentieth century further exacerbated that isolation. The civil wars between the warlords and the central government and between the Nationalists and Communists, the devastating invasions of the 1930s and 1940s by the Japanese, and the prolonged period of China's isolation during Mao's rule further isolated China.

Yet, one of the fascinating aspects of China's emergence over the past three decades has been its efforts to learn from the external world. This has not represented a blatant aping nor an effort to cherry pick ideas from history or Western theoretical writings on strategy and war, but rather a contentious, open debate to examine and draw lessons from West's experience. Two historical case studies have resonated with the Chinese: the Soviet Union's collapse and the rise of Germany in the late nineteenth and early twentieth centuries. These case studies, written in a series of books, were also made into documentary films and form one of the most popular shows on Chinese television.

In the case of the Soviets, the Chinese have drawn the lesson that they must not pursue military development at the expense of economic development – no traditional arms race. That is the path Deng laid out in the late 1970s and one which they have assiduously followed. Indeed, if one examines their emerging military capabilities in intelligence, submarines, cyber, and space, one sees an asymmetrical operational approach that is different from Western approaches, one consistent with the classical Chinese strategic thinkers.

There are interesting trends in the People's Liberation Army (PLA). The Party has ceded considerable autonomy to the military, **allowing the PLA's generals and admirals to build a truly professional force, rather than one constantly hobbled by the party's dictates. This has led to a renaissance in military** thinking; one that draws not only from the classics of Chinese writings, but on an extensive examination of Western literature on history, strategy, and war. The internal consensus seems to be that China is not yet strong enough militarily, and needs to become stronger over the long term. But the debate also extends to issues **on China's strategic and operational choices: Should it be offensive or defensive?** Should it have a continental or maritime focus, or a mixture of the two? How can the PLA best protect the nation's **emerging global** interests?

Above all, the Chinese are interested in the strategic and military thinking of the United States. In the year 2000, the PLA had more students **in America's** graduate schools than the U.S. military, giving the Chinese a growing understanding of America and its military. As a potential future military competitor, China would represent a most serious threat to the United States, because the Chinese could understand America and its strengths and weaknesses far better than Americans understand the Chinese. This emphasis is not surprising, given Sun Tzu's **famous aphorism:**

> *Know the enemy and know yourself; in a hundred battles you will never be in peril. When you are ignorant of the enemy, but know yourself, your chances of winning or losing are equal. If ignorant both of your enemy and of yourself, you are certain in every battle to be in peril.*[20]

In the Second World War and the Cold War, victory by the allies was achieved in part by the thorough understanding of their opponents, who remained relatively ignorant of the United States and its strengths. The Chinese are working hard to ensure that if there is a military confrontation with the United States sometime in the future, they will be ready.

In regard to a potential military competition with the United States, what is apparent in Chinese discussions is a deep respect for U.S. military power. There is a sense that in certain areas, such as submarines, space, and cyber warfare, China can compete on a near equal footing with America. One does not devote the significant national treasure required to build nuclear submarines for coastal defense. The emphasis on nuclear submarines and an increasingly global Navy in particular, underlines worries that the U.S. Navy possesses the ability to shut down China's energy imports of oil – 80% of which go through the straits of Malacca. As one Chinese naval strategist expressed it: *"the straits of Malacca are akin to breathing itself -- to life itself."*

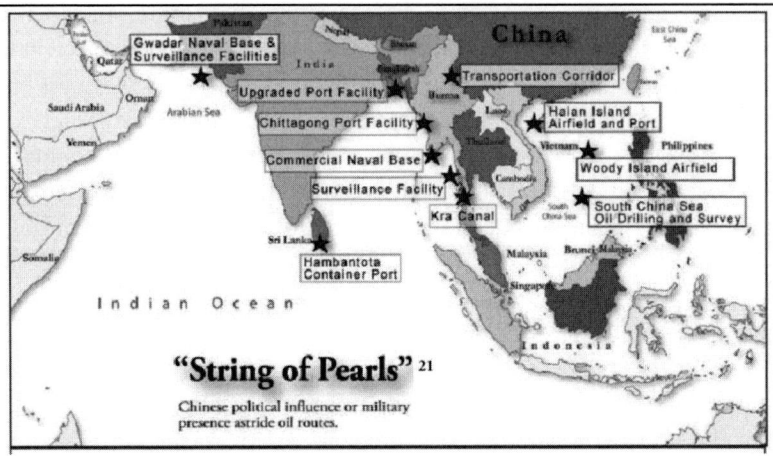

"String of Pearls" 21
Chinese political influence or military presence astride oil routes.

THINKING ABOUT CHINA'S POTENTIAL MILITARY POWER

If GDP alone directly translated into military power, in the 2030s China would have the capacity to afford military forces equal or superior to current U.S. capabilities. And while one must temper such calculations by per capita measures, even by conservative calculations it is easily possible that by the 2030s China could modernize its military to reach a level of approximately one quarter of current U.S. capabilities without any significant impact on its economy. There are some important historical excursions to keep in mind.

First, throughout the Cold War the United States sustained military spending levels, as a percentage of GDP, at about twice current levels, or roughly 8% of GDP, without damaging the economy. If China increased spending to the same level that the United States maintained for decades during the Cold War (8% of GDP) and if U.S. defense spending remained steady as a percent of GDP, China would spend an amount equal to roughly half of America's outlays for defense.

During that period, the Soviet Union proved that a nation can maintain substantially higher rates of military spending for some time before serious economic consequences ensue – the Soviet Union's collapse was due more to the nature of its economic system than to defense expenditures. A similar

> effort by China could see the Chinese equal U.S. defense expenditures for a multi-decade period. Such an effort would quickly come to the attention of Western analysts, but to what effect? Historically, a more obvious massive military buildup such as that taken by Nazi Germany in the years before the Second World War, did not incite the Western powers to respond.

Chinese writers on military and strategic subjects seem to be in agreement that there is a window of opportunity that will last out to the 2020s, during which China can focus on domestic economic growth and expanded trade with the world to make it a truly great power. China is investing significantly in human and physical capital. Indeed, skilled Chinese engineers, technicians, and scientists are deeply involved in scientific discovery around the world, and in building the infrastructure upon which its future prosperity and global integration might be built. Above all, however, the Chinese are objective about their own weaknesses as well as strengths and prospects for the future.

Source: Deutsche Bank

What then are the potential courses that China might follow? The challenges that Chinese leadership confronts at present are enormous, and an unsuccessful China is perhaps more worrisome than a prosperous one. A serious global economic down turn might force China in dangerous directions, as was the case with the Japanese in the 1930s. On the other hand, China is confronting major internal problems that could have an impact on its strategic course. Urbanization, pollution on a monumental scale, water shortages, and the possible responsibility to protect a growing ethnic diaspora, in places such as Siberia or Indonesia, represent realities the leadership cannot easily dismiss. Over the course of its history, internal stability along with the threat of foreign invasions have represented the twin political and strategic challenges that Chinese governments have confronted. Moreover, as recent events in Tibet suggest, tensions between the minorities and the central government in Beijing have been building. Yet with China's approach to strategy, **there is considerable sophistication in the leadership's understanding of its internal problems.**

Taiwan is a wild card, but even here the picture is not clear. A reunification might bring with it the spread of democratic ideals to the mainland and a **weakening of the Party's grip on an increasingly educated and sophisticated** population.

Source: Washington Times

Russian tanks in Georgia

2. Russia

Russia's future remains as uncertain as its past has been tragic. From one of the world's most populous nations with a bright future in 1914, given its natural resources and rapid growth, the world has watched that potential dissipate and then collapse in the catastrophes of World War I (3-4 million military and civilian dead), civil war (5-8 million), man-made famines (6-7 million), purges (2-3 million), and World War II (27 million), accompanied by **sixty years of "planned"** economic and agricultural disasters. The 1990 implosion of the Soviet Union marked a new low point, one that then-President Vladimir Putin decried as **"the greatest geopolitical catastrophe of the century."**

With the collapse of the Soviet Union, Russia lost the lands and territories it had controlled for the better part of three centuries. Not only did the collapse destroy the economic structure that the Soviets created, but the weak structure that the Soviets created, but the weak democratic successor regime proved incapable of controlling the criminal gangs or creating a functioning economy. Moreover, the first attempt by the Russian military to crush the rebellion in Chechnya foundered in a sea of incompetence and faulty assumptions. Since 2000, Russia has displayed a considerable recovery based on Vladimir Putin's **reconstitution of** rule by the security services - a move most Russians have welcomed - and on the influx of foreign exchange **from Russia's production** of petroleum and natural gas. How the Russian government spends that windfall over the long term will play a significant role in the kind of state that emerges.

The nature of the current Russian regime itself also carries significant concerns. To a considerable extent its leaders have emerged from the old KGB. Thus, their education and bureaucratic culture have inculcated them with a ruthlessness that recalls their predecessors, but without their ideological fervor. This suggests that the strategic perspectives of the regime and its zero-sum focus on security bear watching.

At present, Russian leaders appear to have chosen to maximize petroleum revenues without making the long-term investments in oil fields that would increase oil and gas production over the long term. With its riches in oil and gas, Russia is in a position to modernize and repair its ancient and dilapidated infrastructure and improve the welfare of its long suffering people. Nevertheless, the current leadership has displayed little interest in such a course. Instead, it has **placed its emphasis on Russia's great power** status. For all its current riches, the brilliance of Moscow's **resurgence, and the trappings of military** power, Russia cannot hide the conditions of the remainder of the country. The life expectancy of Russia's male population, 59 years, is 148th in the world and places the country somewhere between East Timor and Haiti.

Russian Claimed Territory in Arctic Ocean

Source: University of Durham, UN Marum

Perhaps more than any other nation Russia has reason to fear the international environment, especially considering the invasions that have washed over its lands. There are serious problems: in the Caucasus with terrorists; in Central Asia where the stability of the new oil-rich nations is seriously in question; and in the east where the Chinese remain silent, but increasingly powerful, on the borders of eastern Siberia. In 2001, Russia and China agreed to demarcate the 4,300 mile border between them. However, demographic pressures across this border are increasingly tense as ethnic Russians leave (perhaps as many as a half-million in the 2000-2010 time frame, or 6% of the total population) and ethnic Chinese immigrate to the region. Estimates of the number of ethnic Chinese in Siberia range from a low of about 480,000 (or less than six percent of the population) to more than 1 million (or nearly 12%). Russia must carefully manage this demographic transition to avoid ethnic tensions that could erupt into a cross border conflict with China.

Russia is playing a more active, but less constructive role across the Black Sea, Caucasus, and Baltic regions. Russian involvement in each of these areas has its own character, but they have in common a Russia that is inserting itself into the affairs of its much-smaller neighbors. In each, Russia plays on ethnic and national tension **to extend its influence in its "near abroad."**

In the Caucasus region, especially Georgia and its Abkhazian and South Ossetian provinces, Russia has provided direct support to separatists. In other cases, such as the conflict between Armenia and Azerbaijan or in the Trans-Dnestrian region of Moldova, Russia provides indirect support to keep these conflicts simmering. These conflicts further impoverish areas in dire need of investment and productive economic activity. They lay astride new and vulnerable routes to access the oil of the Caspian Basin and beyond. They encourage corruption, organized crime, and disregard legal order and national sovereignty in a critical part of the world. In the future, they could exacerbate the establishment of frameworks for regional **order and create a new "frontier of** instability" **around** Russia.

Indeed, while many of its European neighbors have almost completely disarmed, the Russians have begun a military buildup, in part to redress the desperately lean years of the 1990s, when the collapse of the post-Soviet economy devastated its military forces. Since 2001, they have quadrupled their military budget with increases of over 20% per annum over the past several years. In 2007, the Russian parliament, with Putin's enthusiastic support, approved even greater military appropriations through 2015. Russia cannot recreate the military machine of the old Soviet Union, but it may be attempting to make up for demographic and conventional military inferiority by modernizing its nuclear forces, including warheads, delivery systems, and doctrines. It is also exploring and fielding strategic systems based on what it terms "new physical principles" including novel stealth and hypersonic technologies. With their vast and increasingly capable nuclear arsenal, the Russians remain a superpower in nuclear terms, despite their demographic and political difficulties.

One of the potential Russias that could emerge in coming decades could be one that focuses on regaining its former provinces in the name of "freeing" the Russian minorities in those border states from the ill-treatment they are supposedly receiving. The United States and its NATO allies would then confront the challenge of summoning up sufficient resolve and deterrence to warn such a Russia off.

At present there is a dangerous combination of paranoia -some of it justified considering Russia's history -nationalism, and bitterness at the loss of what many Russians regard as their rightful place as a great power. It was just such a mixture, along with a series of unfortunate events that drove Nazi Germany on its ill-thought-out course.

3. The Pacific and Indian Oceans

The rim of the great Asian continent is already home to five nuclear powers: China, India, Pakistan, North Korea, and Russia. Furthermore, there are three threshold nuclear states, South Korea, Taiwan, and Japan, which have the capacity to become nuclear powers quickly. While the region appears stable on the surface, political clefts exist. There are few signs that these divisions, which have deep historical, cultural, and religious roots, will be mitigated. China and Korea hold grudges against Japan. Neither China nor Japan have forgotten the seizure of what they regard as their legitimate territory by the Russians. If one includes the breakup of the British Raj in 1947-1948, India and Pakistan have fought three brutal wars, while a simmering conflict over Kashmir continues to poison relations between the two powers. The Vietnamese and the Chinese have a long record of antipathy, which broke out into heavy fighting in the late 1970s. And

China's claim that Taiwan is a province of the mainland obviously represents a flashpoint.

Geographically, there are a number of areas in dispute. The continuing dispute between India and Pakistan over Kashmir is the most dangerous, in this case between two nuclear armed powers. The Chinese have backed up their claims to the Spratleys, which Vietnam and the Philippines also claim, with force. The Kurile Islands, occupied by the Soviets at the end of World War II, remain a contentious issue between Russia and Japan. The uninhabited islands south of Okinawa are in dispute between Japan and China, both obviously drawn to the area by the possibility of oil. Much of the Yellow Sea remains in dispute between the Koreas, Japan, and China, again because of its potential for oil. The straits of Malacca represent the most important transit point for world commerce, the closure of which for even a relatively short period of time would have a devastating impact on the global economy.

There is at present a subtle, but sustained military buildup throughout the region. India could more than quadruple its wealth over the course of the next two decades, but large swaths of its population will likely remain in poverty through the 2030s. Like China, this will create tensions between the rich and the poor. Such tension, added to the divides among its religions and nationalities, could continue to have implications for economic growth and national security. Nevertheless, its military will receive substantial upgrades in the coming years. That fact, combined with its proud martial traditions and strategic location in the Indian Ocean, will make India the dominant player in South Asia and the Middle East. Like India, China and Japan are also investing heavily in military force modernization, particularly with an emphasis in naval forces that can challenge their neighbors for dominance in the seas surrounding the East and South Asian periphery. The buildup of the navies by the powers in the region has significant implications for how the United States develops its strategy as well as for the deployments of its naval forces.

4. Europe

The European Union has solidified Europe economically to a degree not seen since the Roman Empire. For the next quarter century, Europe will exercise **considerable clout in economic matters. The Union's economy as a whole by the 2030s will likely be greater than that of the United States.** From a security standpoint, the NATO alliance will have the potential to field substantial, world-class military forces and project them far beyond the boundaries of the continent, but this is currently a relatively unlikely possibility.

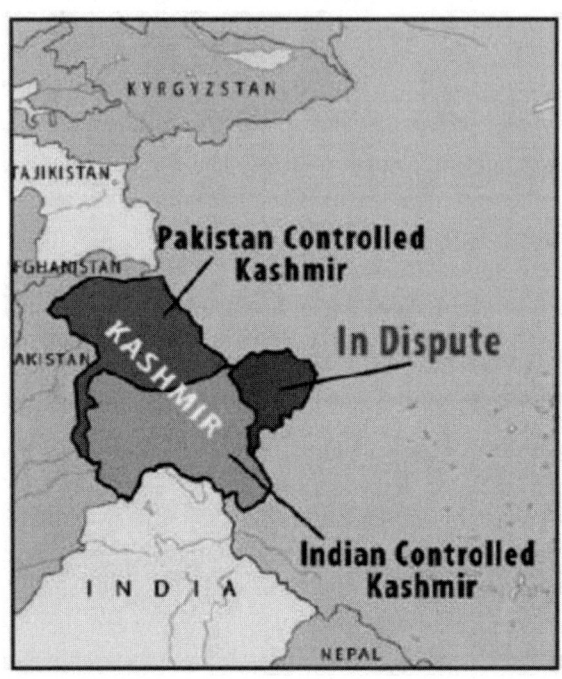

The massive disarmament that occasioned Europe's shift to a "post-conflict" mindset as a reaction to the end of the Cold War will eventually halt, but many European nations have already largely disarmed. The fact that at present only few Europeans have been willing to place their forces in harm's way in support of the NATO commitment in Afghanistan to any significant extent indicates that many Europeans question the idea that lethal military force has a significant role to play in international affairs

Perhaps this will change with the recognition of a perceived threat. The next 25 years will provide two good candidates: Russia and continued terrorism fueled by global Islamic extremism. Russia has already been discussed. The Baltic and Eastern European regions will likely remain flashpoints as a number of historical issues such as ethnicity or the location of national boundaries, that have in the past led to conflict, continue to simmer under the surface. Russian efforts to place the gas pipeline to Western Europe through the Baltic rather than through Eastern Europe suggests a deliberate aim to separate the Central and Western European NATO countries from the Baltic and Eastern European members of NATO.

Continued terrorist attacks in Europe might also spark a popular passion for investing in military forces. Should violent extremists persist in using this tactic to attack the European continent with increasing frequency and intensity, there might a response that includes addressing this threat on a global scale rather than as an internal security problem.

5. Central and South America

The military problems that arise in South America and Central America will likely arise from within. Many currently plague the continent, particularly drug cartels and criminal gangs, while terrorist organizations will continue to find a home in some of the continent's lawless border regions.

Nevertheless, South America's improving economic situation suggests the region could be in a better position to deal with these problems. Brazil, in particular, appears set on a course that could make it a major player among the great powers by the 2030s. Chile, Argentina, Peru and possibly Colombia will also most likely see sustained growth, if they continue prudent economic policies.

The potential major challenges to the status quo at present are Cuba and Venezuela. The demise of the Castros will create the possibility of major changes in Cuba's politics. The future of Venezuela is harder to read. The Chavez regime is diverting substantial amounts of its oil revenues to further its anti-American "Bolivarian Revolution," while at the same time consolidating his regime's hold on power by distributing oil wealth to his supporters. By trying to do both it is shortchanging investments in its oil infrastructure, which has serious implications for the future. Unless its current regime changes direction, it could use its oil wealth to subvert its neighbors for an extended period while pursuing anti-American activities on a global scale with the likes of Iran, Russia, and China, in effect creating opportunities to form anti-American coalitions in the region.

Brazil will become a superpower in regional terms. No country in South America is likely to approach its economic power, which will rapidly grow stronger due to its resources in biofuels. The oil fields that have been found off in the Brazilian coast represent a resource that will add to Brazil's economic and political power.

A serious impediment to growth in Latin America remains the power of criminal gangs and drug cartels to corrupt, distort, and damage the region's potential. The fact that criminal organizations and cartels are capable of building dozens of disposable submarines in the jungle and then using them to smuggle cocaine, indicates the enormous economic scale of this activity. This posesareal threat to the national security interests of the Western Hemisphere. In particular, the growing assault by the drug cartels and their thugs on the Mexican government over the past several years reminds one that an unstable Mexico could represent a homeland security problem of immense proportions to the United States.

Source: Department of Defense

6. Africa

Sub-Saharan Africa presents a unique set of challenges, including bad governance, interference by external powers, and health crises such as AIDS. Even pockets of economic growth are under pressure and may soon regress. Some progress in the region may occur, but it is almost certain that many of these nations will remain on any list of the poorest nations on the globe. Exacerbating their difficulties will be the fact that the national borders, drawn by the colonial powers in the nineteenth century, bear little relation to tribal and linguistic realities.

The region is endowed with a great wealth of natural resources, a fact which has already attracted the attention of several powerful states. This could represent a welcome development, because in its wake could come foreign expertise and investment for a region in dire need of both. The importance of the region's resources will ensure the great powers maintain a vested interest in the region's stability and development. If this engagement goes beyond "aid" to become true "investment," then true stability and security may emerge. Until that happens, the main driver for joint force involvement in Africa will be to avert humanitarian and genocidal disasters as African states and sub-state tribal groups struggle for power and control among themselves. Relatively weak African states will be very hard-pressed to resist pressure by powerful state and non-actors who embark on a course of interference. This possibility is reminiscent of the late nineteenth century, when pursuit of resources and areas of interest by the developed world disturbed the affairs of weak and poverty stricken regions.

7. *The Center of Instability: The Middle East and Central Asia*

On current evidence, a principal nexus of conflict will continue to be the region from Morocco to Pakistan through to Central Asia. Across this part of the globe exist a number of historical, dormant conflicts between states and nations over borders, territories, and water rights, especially in Central Asia and the Caucasus. Radical Islamists will present the first and most obvious challenge. The issue here is not terrorism per se, because terrorism is merely a tactic by which those who lack the technology, weapons systems, and scruples of the modern world can attack their enemies throughout the world. Radical Islamists who advocate violence – and not all do – constitute a transnational, theologically-based insurgency which seeks to overthrow regimes in the Islamic world. They bitterly attack the trappings of modernity as well as the philosophical underpinnings of the West despite the fact their operations could not be conducted without the internet, air travel and globalized financial systems. At a minimum radical Islam seeks to eliminate U.S. and other foreign presence in the Middle East, a region vital to U.S. and global security, but only as a first step to the creation of a Caliphate stretching from Central Asia in the East to Spain in the West and extending deeper into Africa, overwhelming Christian and indigenous religions and ensuring that "Islam's bloody borders" remain inflamed.[22]

The problems in the Arab-Islamic world stem from the past five centuries, during which, until recently, the rise of the West and the dissemination of Western political and social values paralleled a concomitant decline in the power and appeal of their societies. Today's Islamic world confronts the choice of either adapting to or escaping from a globe of interdependence created by the West. Often led by despotic rulers, addicted to the exports of commodities which offered little incentive for more extensive industrialization or modernization, and burdened by cultural and ideological obstacles to education and therefore modernization, many Islamic states have fallen far behind the West, South Asia, and East Asia. The rage of radical Islamists feeds off the lies of their often corrupt leaders, the rhetoric of radical imams, the falsifications of their own media, and resentment of the far more prosperous developed world. If tensions between the Islamic world's past and the present were not enough, the Middle East, the Arab heartland of Islam, remains divided by tribal, religious, and political divisions, in which continued instability is inevitable.

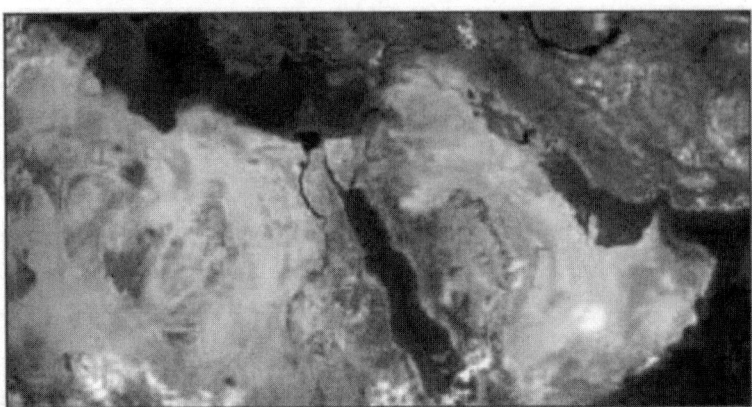

Source: NASA

Combining Islamic dogma with the internet, intricate financial networks, and the porous borders of weakly governed states, radical Islamists have created a networked organization with global reach. The movement is similar to most insurgencies in the fanaticism of its leaders. But the ability to employ advanced technologies with the aim of causing maximum destruction represents a dangerous new trend in the international environment.

No one should harbor the illusion that the developed world can win this conflict in the near future. As is true with most insurgencies, victory will not appear decisive or complete. It will certainly not rest on military successes. The treating of political, social, and economic ills can help, but in the end will not be decisive. What will matter most will be the winning of a "war of ideas," much of which must come from within the Islamic world itself.

The economic importance of the Middle East with its energy supplies hardly needs emphasis. Whatever the outcome of the conflicts in Iraq and Afghanistan, U.S. forces will find themselves again employed in the region on numerous missions ranging from regular and irregular war, relief and reconstruction, to engagement operations. The region and its energy supplies are too important for the U.S., China, and other energy importers to allow radical groups to gain dominance or control over any significant portion of the region.

C. Weak and Failing states

Weak and failing states will remain a condition of the global environment over the next quarter of a century. Such countries will continue to present strategic and operational planners serious challenges, with human suffering on a scale so large that it almost invariably spreads throughout the region, and in some cases possesses the potential to project trouble throughout the globalized world.

Yet, there is no clear pattern for the economic and political troubles that beset these states. In some cases, disastrous leadership has wrecked political and economic stability. In others, wars among tribal groups with few cultural, linguistic, or even racial ties have imploded states. This was the case in Africa and the Middle East, where in the nineteenth century the European powers divided frontiers between their colonies on the basis of economic, political, or strategic necessity and paid scant attention to existing linguistic, racial, or cultural patterns of the tribal societies. These dysfunctional borders have exacerbated nearly every conflict in which our forces have been involved in these regions.

Many, if not the majority, of weak and failing will center in Sub-Saharan Africa, Central Asia, the Middle East, and North Africa. A current list of such states much resembles the lists of such states drawn up a generation ago, suggesting a chronic condition, which, despite considerable aid, provides little hope for solution. There have been a number of nations that have escaped poverty – their successes resulting from intelligent leadership and a willingness to embrace integration into the global system. To date, the remaining weak and failing nations have chosen other paths.

There is one dynamic in the literature of weak and failing states that has received relatively little attention, **namely the phenomenon of "rapid collapse."** For the most part, weak and failing states represent chronic, long-term problems that allow for management over sustained periods. The collapse of a state usually comes as a surprise, has a rapid onset, and poses acute problems. The collapse of Yugoslavia into a chaotic tangle of warring nationalities in 1990 suggests how suddenly and catastrophically state collapse can happen - in this case, a state which had hosted the 1984 Winter Olympics at Sarajevo, and which then quickly became the epicenter of the ensuing civil war.

In terms of worst-case scenarios for the Joint Force and indeed the world, two large and important states bear consideration for a rapid and sudden collapse: Pakistan and Mexico.

Some forms of collapse in Pakistan would carry with it the likelihood of a sustained violent and bloody civil and sectarian war, an even bigger haven for violent extremists, and the question of what would happen to its nuclear weapons. That **"perfect storm" of uncertainty** alone might require the engagement of U.S. and coalition forces into a situation of immense complexity and danger with no guarantee they could gain control of the weapons and with the real possibility that a nuclear weapon might be used.

The Mexican possibility may seem less likely, but the government, its politicians, police, and judicial infrastructure are all under sustained assault and pressure by criminal gangs and drug cartels. How that internal conflict turns out over the next several years will have a major impact on the stability of the Mexican state. Any descent by Mexico into chaos would demand an American response based on the serious implications for homeland security alone.

D. The Threats of Unconventional Power

While states and other conventional powers will remain the principal brokers of power, there is an undeniable diffusion of power to unconventional, non-state, or trans-state actors. While these groups have "rules" of their own, they exist and behave outside the recognized norms and conventions of society.

Some transnational organizations seek to operate beyond state control and acquire the tools and means to challenge states and utilize terrorism against populations to achieve their aims. These unconventional transnational organizations possess no regard for international borders and agreements. The discussion below highlights two examples: militias and super-empowered individuals.

Militias represent armed groups, irregular yet recognizable as an armed force, operating within ungoverned areas or in weak failing states. They range from ad hoc organizations with shared identities to more permanent groups possessing the ability to provide goods, services, and security along with their military capabilities. Militias challenge the sovereignty of the state by breaking the monopoly on violence traditionally the preserve of states. An example of a modern day militia is Hezbollah, which combines state-like technological and warfighting capabilities with a "substate" political and social structure inside the formal state of Lebanon.

One does not need a militia to wreak havoc. Pervasive information, combined with lower costs for many advanced technologies, has already resulted in individuals and small groups possessing increased ability to cause significant damage and slaughter. Time and distance constraints are no longer in play. Such groups employ niche technologies capable of attacking key systems and providing inexpensive countermeasures to costly systems. Because of their small size, such groups of the "super-empowered" can plan, execute, receive feedback, and modify their actions, all with considerable agility and synchronization. Their capacity to cause serious damage is out of all proportion to their size and resources.

The global effort against terrorist organizations will continue into the 2030s with varying degrees of intensity over time. It will most likely remain at the forefront of U.S. security concerns. At present, the evidence suggests that U.S. efforts have largely decimated the al Qaeda that attacked the United States in 2001. However, the threat has not disappeared, as new radical cadres have formed. These new terrorist groups have learned from al Qaeda's shortcomings and mistakes. Moreover, the ability of terrorist organizations to utilize the internet and other means of communications to pass the experiences, tactics, and best training methods will result in a constant flow of relatively sophisticated new volunteers to the fight. The ability of terrorists to learn from their predecessors and colleagues will not confront the hindrance of having to process adaptations and innovations through bureaucratic barriers.

E. The Proliferation of Weapons of Mass Destruction

A continuing challenge to American security will be the proliferation of nuclear weapons. Throughout the Cold War, U.S. planners had to consider the potential use of nuclear weapons both by and against the Soviet Union. For the past twenty years, Americans have largely ignored issues of deterrence and nuclear warfare. In the 2030s, they will no longer have that luxury.

Since 1998, India and Pakistan have created nuclear arsenals and delivery capabilities. North Korea has tested a nuclear weapon and has produced sufficient fissile material to create more such weapons. At present, the Iranians are pressing forward aggressively with their own nuclear weapons program. The confused reaction in the international community to Iran's defiance of external demands to terminate its nuclear development programs is an incentive for others to follow in their path.

In effect, there is a growing arc of nuclear powers running from Israel in the west through an emerging Iran to Pakistan, India, and on to China, North Korea, and Russia in the east. Both Taiwan and Japan have the capability to develop nuclear weapons quickly, should their political leaders decide to do so. Unfortunately, that nuclear arc coincides with areas of considerable instability - regions that because of their economic power and energy resources are of enormous interest to the United States.

Moreover, some in the region might not view nuclear weapons as weapons of last resort. It is far from certain that a state whose culture is deeply distinct from that of the United States, and whose regime is either unstable or unremittingly hostile (or both), would view the role of nuclear weapons in a fashion similar to

American strategists. The acquisition of nuclear weapons by other regimes throughout this arc, whether they were hostile or not, would disrupt the strategic balance further, while increasing the potential for the use of nuclear weapons. Add to this regional complexity the fact that multiple nuclear powers will very likely have the global reach to strike other states around the world. The stability of relations among numerous states capable of global nuclear strikes will be of central importance for the Joint Force. Assured second-strike capabilities and relations based on mutually-assured destruction may mean greater stability, but may effectively reduce access to parts of the world. On the other hand, fragile nuclear balances and vulnerable nuclear forces may provide tempting targets for nuclear armed competitors.

Any discussion of weapons of mass destruction must also address the potential use of biological weapons by sovereign states as well as non-state actors. By all accounts, such weapons are becoming easier to fabricate – certainly easierthannuclearweapons –and under the right conditions they could produce mass casualties, economic disruption, and terror on the scale of a nuclear strike. The knowledge associated with developing biological weapons is widely available, and the costs for their production remain modest, easily within reach of small groups or even individuals.

Source: Korean Central News Agency

North Korea flaunts its strategic might

F. Technology

Advances in technology will continue at an exponential pace as they have over the past several decades. Some pundits have voiced worries the United States will lose its lead as the global innovator in technology or that an enemy could make technological leaps that would give it significant advantages militarily. That is possible, but by no means a foregone conclusion.

It is clear that technology, distinct from weapons of mass destruction (WMD), will proliferate. As anyone who has purchased a home computer knows, technological advances drive down the overall cost of ever-greater capability. The weapons market is no different. More advanced weaponry will be available to more groups, conventional and unconventional, for a cheaper price. This will allow relatively moderately funded states and militias to acquire long-range precision munitions, projecting power farther out, and with greater accuracy, than ever before. At the high end, it has already been seen that this reach extends into space with the public demonstration of anti-satellite weapons. Furthermore, the market for advanced weaponry potentially empowers any small actor or group, as long as they have the cash. Whether a small oil-rich nation or a drug cartel, cash will be able to purchase lethal capabilities. If manpower is a limiting factor, the advances in robotics provide a solution for those who can afford the price. This **has the sobering potential to further amplify the power of the "super-**empowered **guerilla."**

In the globalized, connected world of science and technology, there is less chance that major technological advances could catch American scientists by surprise. In the past, the real issue with technology has not been simply that a particular nation has developed weapons far superior to those of its opponents. Rather, in nearly every case the major factor has been how military organizations have integrated technological advances into their doctrinal and tactical system. It has been the success or failure in that regard that has resulted in battlefield surprise and success. In 1940 French tanks were superior in almost every respect to those of the Germans. What gave the Wehrmacht its unique advantage was the fact that the Germans integrated the tank into a combined arms team. The real surprise of Blitzkrieg lay in the inability of the French to imagine how the Germans might exploit battlefield success with the new technologies available. It was the development of decentralized, combined-arms tactics by the Germans that led to their overwhelming victory, not new, more sophisticated weapons systems.

Thus, what has been unquestionably crucial is the degree of imagination military organizations have displayed in incorporating new technologies into their doctrine and concepts. The fact that the speed of technological change and invention proceeds exponentially will make the ability to adapt new technologies into the larger framework of military capability even more critical in coming decades.

A current example of the kind of technological surprise that could prove deadly would be an adversary's deployment and use of a disruptive technology, such as electro-magnetic pulse (EMP) weapons against a force without properly hardened equipment. The potential effects of an electro-magnetic pulse resulting from a nuclear detonation has been known for decades. The appearance of non-nuclear EMP weapons could change operational and technological equations. They are being developed, but are joint forces being adequately prepared to handle such a threat? The impact of such weapons would carry with it the most serious potential consequences for the communications, reconnaissance, and computer systems on which the Joint Force depends at every level.

TECHNOLOGY, DOCTRINE, AND SUCCESSFUL ADAPTATION

Nothing more clearly illustrates the importance of imagination and an understanding of war in the incorporation of technology into military institutions than the utilization of radar over the course of the first two years of World War II. It was not until the 1930s that scientists in the major powers turned their attention to the possibility radio waves could spot the flight of aircraft or the movement of ships at sea. The looming threat posed by enemy bombers in a period of worsening international tensions instigated investigations into such possibilities. By the late 1930s scientists in Britain, Nazi Germany, and the United States had all developed workable capabilities for identifying the height, direction, and speed of aircraft, as well as the number of aircraft.

Not surprisingly, the Germans, given their technological prowess, developed the most sophisticated radars, but the incorporation of that technological capability into their weapons systems lagged behind that of the British. It would be in the Battle of Britain that German failure in imagination would show to the greatest extent. The Luftwaffe had incorporated radar into its capabilities in the late 1930s, but only as a series of ground control intercept sites, each of which operated independently with no direct tie to a larger air defense system. It would not be until the catastrophe of Hamburg in summer 1943 that the Luftwaffe would create an air defense system in which radar formed in integral part of a holistic approach to an air warning system of

> defense. But the British were already using such a system in 1940. As the scientific intelligence officer, R.V. Jones, recalled in his memoirs:
>
> > *[The] German philosophy of [air defense] ran roughly along the lines that here was an equipment which was marvelous in the sense that it would enable a single station to cover a circle of a radius 150 kilometers and detect every aircraft within that range.... Where we had realized that in order to make maximum use of radar information the stations had to be backed by a communications network which could handle the information with the necessary speed, the Germans seemed simply to have grafted their radar stations on to the existing observer corps network which had neither the speed nor the handling **capacity that the radar information merited**.... The British approach... was entirely different. The British radar stations formed the eyes of a systematic approach to the air defense of the British Isles, so that RAF commanders could use their information to guidelarge numbers of Hurricanes and Spitfires against German bomber formations.*[23]
>
> As Churchill noted in his memoirs about the Second World War, "it was the operational efficiency rather than novelty of equipment that was the British achievement."[24]

Finally, it is by no means certain that the United States and its allies will maintain their overall lead in technological development over the next 25 years. **America's secondary** educational system is clearly declining in a relative sense when compared to leading technological competitors, for instance India and China. **America's** post-graduate educational programs and research laboratories **remain the world's most advanced** -magnets for some of the best scientific minds in the world. However, although many foreign students remain in the United States, significant numbers are now returning home. Without substantive changes to improve its educational system, the United States will pay a heavy price in the future.

G. The Battle of Narratives

Modern wars are fought in more than simply the physical elements of the battlefield. Among the most important of these are the media **in which "the battle to win the narrative" will occur. Our enemies** have already recognized that perception is as important to their success as the actual event. For terrorists, the internet and mass media have become forums for achieving their strategic and political aims. Sophisticated terrorists emphasize the importance of integrating combat activities (terrorist attacks) into a coherent strategic communications program. Radical groups are not the only ones who understand the importance of dominating the media message. A major state synchronizing military operations with a media offensive was on display during Russia's invasion of Georgia. Within days of the invasion, a small coterie of Russians, well known in the West, was placing editorials in major newspapers in the United States and Europe.

The battle of the narrative must involve a sophisticated understanding of the enemy and how he will attempt to influence the perceptions not only of his followers, but the global community. His efforts will involve deception, sophisticated attempts to spin events, and outright lies. As Joseph Goebbels, the evil Minister of Propaganda for the Third Reich, once commented, the bigger the lie, the greater its influence. No matter how outlandish enemy claims may seem to Americans, those charged with the responsibility for information operations must understand how those who will receive the message will understand it. In this regard, they should not forget that the KGB's efforts at the end of the Cold War to persuade Africans that the CIA was responsible for the spread of Acquired Immune Deficiency Syndrome (AIDS) throughout their continent are still reverberating in parts of Africa. Information has been, is, and will continue to be a strategic and political weapon. Its power will only increase as communications technology and the density of global media become more pervasive. At the end of the day, it is the perception of what happened that matters more than what may actually have happened.

Dominating the narrative of any operation, whether military or otherwise, pays enormous dividends. Failure to do so undermines support for policies and operations, and can actually damage a country's reputation and position in the world. For example, in the aftermath of Hurricane Katrina, America's global standing fell sharply, while many Americans remain convinced their government's reaction was at best inept and at worst a reflection of latent racism. In truth, at the end of the first week after the disaster 38,000 federal troops were supporting the National Guard and local authorities. They were already caring for approximately 100,000 displaced citizens, had fed over a million meals, and had provided medical care to tens of thousands.

Compare the reaction to Katrina to the reaction to the nation's previous most destructive storm, Hurricane Andrew. At the end of the first week after Andrew, not a single federal soldier had gone to work, and less than 1,500 had deployed. Yet, the federal government's reaction to Andrew is graded a success, while the much larger, and infinitely more efficient response to Katrina is almost universally judged a failure. The reason for such perceptions lies in the fact that an inept strategic communications operation lost control of the narrative.

In the battle for the narrative, the United States must not ignore its ability to bring its considerable soft power to bear in order to reinforce the positive aspects of joint force operations. Humanitarian assistance, reconstruction, securing the safety of local populations, military-to-military exercises, health care, and disaster relief are just a few examples of the positive measures that we offer. Just as no nation in the world can respond with global military might on the scale of the United States, so too are we unmatched in our capacity to provide help and relief across thousands of miles. All of these tools should be considered in this battle to build trust and confidence.

In the future, influencing the narrative by conveying the truth about **America's intent, reinforced with supporting actions and activities, will become** even harder. As communications technologies become more widely available, an ever-wider array of media will influence global public opinion. The U.S. government and its joint forces will always be held to a much higher standard in this area than our adversaries. Joint force commanders already wrestle with how to deal with a pervasive media presence, widespread blogging, almost instantaneous posting of videos from the battlefield, e-mail, and soldiers who can call home whenever they return to base. In the future they will be confronted with a profusion of new media linked to unimaginably fast transmission capabilities. Just as we have already begun to think of every Soldier and Marine as an intelligence collector, we will also have to start considering them as global communications producers. Today, commanders talk about the strategic corporal, whose acts might have strategic consequences if widely reported. This still remains a hit or miss affair, less often requiring the presence of the media representative to attract a global reaction.

Even in the past, the success of combat operations has not always been judged on the battlefield. In 1968 the Tet Offensive was smashed by the American military, but the narrative reported in the United States served to undermine support for the war effort. U.S. weapons employment in this battle of the narratives must be in consonance with the message, even if it means sometimes bypassing tactical targets. Winning the battle has always been important, but in the pervasive and instantaneous communications environment expected in future decades, it will be absolutely crucial. For commanders not to recognize that fact could result in the risking of the lives of young Americans to no purpose.

Source: U.K. DCDC

Splendor amid squalor: Modern Rio de Janeiro

H. Urbanization

By the 2030s, five of the world's eight billion people will live in cities. Fully two billion of them will inhabit the great urban slums of the Middle East, Africa, and Asia. Moreover, while at present half of the world's poorest 10% live in Asia, that share will decrease to one fifth, while Africa's will rise from one–third to two–thirds. Most mega-cities and cities will lie along the coast or in littoral environments. With so much of the world's population crammed into dense urban areas and their immediate surroundings, future joint force commanders will be unable to evade operations in urban terrain. The world's cities with their teeming populations and slums will be places of immense confusion and complexity, physically as well as culturally. They also will provide prime locations for diseases and the population density for pandemics to spread.

There is no historical precedent for major cities collapsing, even in the eighteenth and nineteenth centuries, when the first such cities appeared. Cities under enormous stress, such as Beirut in the 1980s and Sarajevo in the 1990s, nevertheless managed to survive with only brief interruptions of food imports and basic services. As in World War II, unless contested by an organized enemy, urban areas are always easier to control than the countryside. In part, that is because cities offer a pre–existing administrative infrastructure through which forces can manage secured areas while conducting stability operations in contested locations. The effectiveness of that pre-existing infrastructure may be tested as never before under the stress of massive immigration, energy demand, and food and water stress in the urban sprawl that is likely to emerge. More than ever before, it will demand the cultural and political knowledge to utilize that infrastructure.

Urban operations will inevitably require the balancing of the disruptive and destructive military operations with the requirements of humanitarian, security, and relief and reconstruction operations. What may be militarily effective may also create the potential for large civilian casualties, which in turn would most probably result in a political disaster, especially given the ubiquitous presence of the media.

As well, the nature of operations in urban environments places a premium on decentralized command and control, ISR, fire support, and aviation. Combat leaders will need to continue to decentralize decision-making down to the level where tactical leaders can act independently in response to fleeting opportunities.

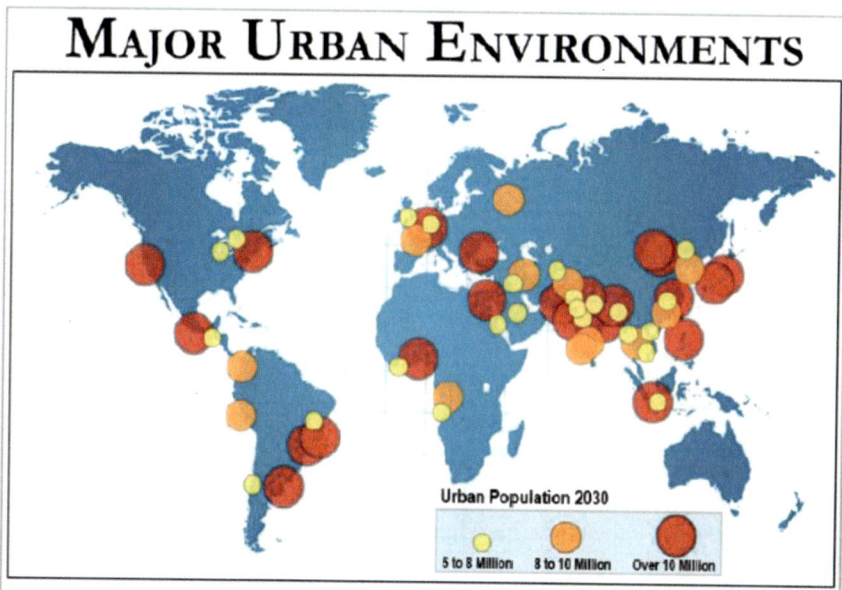

- In 2007, 50% of the world's population lived in cities.
- By 2030, 65% of humanity will live in cities-5.5 billion human beings.
- Megacities are located near the oceans, and subject to severe environmental, social, and political pressures.

Source: United Nations World Urbanization Prospects

PART IV: THE IMPLICATIONS FOR THE JOINT FORCE

> *Order or disorder depends on organization; courage or cowardice on circumstances; strength or weakness on dispositions…Thus, those skilled at making the enemy move do so by creating a situation to which he must conform; they entice him with something he is certain to take, and with lures of ostensible profit they await him in strength. Therefore a skilled commander seeks victory from the situation and does not demand it of his subordinates.*[25]
>
> Sun Tzu

In an uncertain world, which will inevitably contain enemies who aim to either attack the United States directly or to undermine the political and economic stability on which America, its allies, and the world's economy depend, the nation's military forces will play a crucial role. Yet, war is an inherently uncertain and costly endeavor. As the United States has discovered in Iraq and Afghanistan, there is no such thing as a rapid, decisive operation that does not generate unforeseen second and third order effects.

While the most important mission of the American military has been the ability to fight and win the nation's wars, the ability of U.S. forces to deter conflict has risen to equal footing. Preventing war will prove as important as winning a war. In fact, the two missions are directly linked in a symbiotic relationship. The ability to deter a potential adversary depends on the capabilities and effectiveness of U.S. forces to act across the full range of military operations. Deterrence also depends on the belief on the part of the adversary that the United States will use its military power in defense of its national interests.

Since the fall of the Iron Curtain, the United States has planned for a global repositioning effort, removing forces from forward basing and garrisoning much of its military force structure at home. Instead, the Joint Force has found itself in near-constant conflict abroad, and now forces based at home find themselves in heavy rotation, projecting forward into the Middle East and elsewhere around the world. After protracted action in Afghanistan and Iraq, the force now faces a period of reconstitution and rebalancing which will require significant physical, intellectual, and moral effort that may take a decade to complete. During this time, our forces may be located significant distances from a future fight. Thus, the Joint Force will be challenged to maintain both a deterrent posture and the capacity and capability to be forward engaged around the world, showing the flag and displaying the ability to act in ways to both prevent and win wars.

A. War in the Twenty-First Century

As the discussion of trends and contexts above has suggested, the roles and missions of the Joint Force will include the protection of the homeland, the maintenance of the global commons, the deterrence of potential enemies, and, when necessary, fighting and winning conflicts that may occur around the world. Such challenges are by themselves daunting enough, but they will occur in a period characterized by radical technological, strategic, and economic change, all of which will add to the complexities of the international environment and the use of military **force. America's position in the world, unprecedented in almost every** respect, will continue to present immense challenges to its military forces.

Rapidly changing trends within the contexts described in the previous section will have profound implications for the character of war itself and the methods by which the Joint Force will wage it. Yet, the nature of war will remain closer to Agincourt than to Star Trek. At its heart, war will always involve a battle between two creative human forces. Our enemies are always learning and adapting. They will not approach conflicts with conceptions or understanding similar to ours. And they will surprise us. No amount of technology, conceptualization, or globalization will change those realities. Moreover, the employment of military force will continue to be conditioned by politics --not only those of the United States and its allies, but by those of its opponents. Above all, joint force commanders, their staffs, and their subordinates must have a clear understanding of the strategic and political goals for which they conduct military operations. In almost every case, they will find themselves working closely with partners, a factor which will demand not only a thorough understanding of U.S. political goals, but coalition goals as well.

It is in this political-strategic environment that the greatest surprises for Americans may come. The United States has dominated the world economically since 1915 and militarily since 1943. Its dominance in both respects now faces challenges brought about by the rise of powerful states. Moreover, the rise of these great powers creates a strategic landscape and international system, which, despite continuing economic integration, will possess considerable instabilities. Lacking either a dominant power or an informal organizing framework, such a system will tend toward conflict. Where and how those instabilities will manifest themselves remains obscure and uncertain.

Between now and the 2030s, the military forces of the United States will almost certainly find themselves involved in combat. Such involvement could come in the form of a major regular conflict or in a series of wars against insurgencies. And as this document has suggested, they will certainly find themselves engaged not only against terrorist organizations, but against those who sponsor them. One of the great problems that confronts American strategists and military planners is the conundrum of preparing for wars that remain uncertain as to their form, location, level of commitment, the contribution of potential allies, and the nature of the enemy. The only matter that is certain is that joint forces will find themselves committed to conflict against the enemies of the United States and its Allies, and in defense of its vital interests.

B. Preparing for War

There are two ominous scenarios that confront joint forces between now and the 2030s. The first and most devastating would be a major war with a powerful state or hostile alliance of states. Given the proliferation of nuclear weapons, there is the considerable potential for such a conflict to involve the use of such weapons. While major regular war is currently in a state of hibernation, one should not forget that in 1929 the British government adopted as its basic principle of defense planning the assumption that no major war would occur for the next ten years. Until the mid-1930s "the ten year rule" crippled British defense expenditures. The possibility of war remained inconceivable to British statesmen until March 1939.

The one approach that would deter a major conflict involving U.S. military forces, including a conflict involving nuclear weapons, is the maintenance of capabilities that would allow the United States to wage and win any possible conflict. As the Romans so aptly commented, "if you wish for peace, prepare for war." Preventing war will in most instances prove more important than waging it. In the long-term, the primary purpose of the military forces of the United States must be deterrence, for war in any form and in any context is an immensely expensive undertaking both in lives and national treasure. When, however, deterrence fails, then, the military effectiveness of those forces will prove crucial. Here the efforts that have gone into preparing U.S. forces for conflict at their various training centers must continue to receive the same support and attention in the future that they have over the course of the past 30 years. As the Japanese warrior/commentator Miyamoto Musashi noted in the seventeenth century:

> *There is a rhythm in everything, but the rhythms of the art of war are especially difficult to master without practice.... In battle, the way to win is to know the opponent's rhythms while using unexpected rhythms yourself, producing formless rhythms from the rhythms of wisdom.*[26]

The second ominous scenario that confronts the Joint Force is the failure to recognize and fully confront the irregular fight that we are in. The requirement to prepare to meet a wide range of threats is going to prove particularly difficult for American forces in the period between now and the 2030s. The difficulties involved in training to meet regular and nuclear threats must not push preparations to fight irregular war into the background, as occurred in the decades after the Vietnam War. Above all, Americans must not allow themselves to be deluded into believing their future opponents will prove as inept and incompetent as Saddam Hussein's regime was in 1991 and again in 2003. Having seen the capabilities of U.S. forces in both regular and irregular war, future opponents will understand "the American way of war" in a particularly detailed and thorough way.

In Iraq and Afghanistan our opponents have displayed considerable capacity to learn and adapt in both the political and tactical arenas. More sophisticated opponents of U.S. military forces will certainly attack American vulnerabilities. For instance, it is entirely possible that attacks on computers, space, and communications systems will severely degrade command and control of U.S. forces. Thus, those forces must possess the ability to operate effectively in degraded conditions.

In planning for future conflicts, joint force commanders and their planners must factor two important constraints into their calculations: logistics and access. The majority of America's military forces will find themselves largely based in North America. Thus, the first set of problems involved in the commitment of U.S. forces will be logistical. In the 1980s many defense pundits criticized the American military for its supposed over-emphasis on logistics, and praised the German Wehrmacht for its minimal "tooth to tail" ratio in the Second World War. What they missed was that the United States had to project its military forces across two great oceans, then fight massive battles of attrition in Europe and in East Asia. Ultimately, the logistical prowess of U.S. and Allied forces, translated into effective combat forces, defeated the Wehrmacht on the Western Front, crushed the Luftwaffe in the skies over Germany, and broke Imperial Japan's power.

The tyranny of distance will always influence the conduct of America's wars, and joint forces will confront the problems associated with moving forces over great distances and then supplying them with fuel, munitions, repair parts, and sustenance. In this regard, a measure of excess is always necessary, compared to "just in time" delivery. Failure to keep joint forces who are engaged in combat supplied could lead to disaster, not just unstocked shelves. Understanding that requirement represents only the first step in planning, but it may well prove the most important.

The crucial enabler for America's ability to project its military power for the past six decades has been its almost complete control over the global commons. From the American standpoint, the Battle of the Atlantic that saw the defeat of the German U-boat menace in May 1943 was the most important victory of the Second World War. Any projection of military power in the future will require a similar enabling effort, and must recognize that the global commons have now expanded to include the domains of cyber and space. The Joint Force must have redundancy built in to each of these areas to ensure that access and logistics support are more than "single-point safe" and cannot be disrupted through a single enemy point of attack.

In America's two recent wars against Iraq, the enemy made no effort to deny U.S. forces entry into the theater. Future opponents, however, may not prove so accommodating. Hence, the second constraint confronting planners is that the United States may not have uncontested access to bases in the immediate area from which it can project military power. Even in the best case, allies will be essential to providing the base structure required for arriving U.S. forces. But there may be other cases where uncontested access to bases is not available for the projection of military forces. This may be because the neighborhood is hostile, or because smaller friendly states have been intimidated. Hence, the ability to seize bases by force from the sea and air could prove the critical opening move of a campaign.

Given the proliferation of sophisticated weapons in the world's arms markets – potential enemies – even relatively small powers will be able to possess and deploy an array of longer-range and more precise weapons. Such capabilities in the hands of America's enemies will obviously threaten the projection of forces into a theater as well as attack the logistical flow on which U.S. forces will depend. Thus, the projection of military power could become hostage to the ability to counter long-range systems even as U.S. forces begin to move into a theater of operations and against an opponent. The battle for access may prove not only the most important, but the most difficult.

One of the major factors in America's success in deterring potential aggressors and projecting its military power over the past half century has been the presence of its naval forces off the coasts of far-off lands. Moreover, those forces have also proven of enormous value in relief missions when natural disasters have struck. They will continue to be a significant factor in the future. Yet, there is also the rising danger with the increase in precision and longer range missiles that presence forces could be the first target of an enemy's action in their exposed positions.

C. The Conduct of Military Operations in the Twenty-First Century

The forms of future war will each present peculiar and intractable challenges to joint forces. The U.S. will always seek to fight and operate with partners, leading where appropriate, and prepared to act alone when required to support our vital national interests. However, there is every likelihood that there will be few lines of delineation between one form of conflict and another. Even in a regular war, potential opponents, engaged in a life and death struggle with the United States, may engage U.S. forces across the spectrum of conflict. Thus, the Joint Force must expect attacks on its sustainment, its intelligence, surveillance and reconnaissance (ISR) capabilities, and its command and control networks. The Joint Force can expect future opponents to launch both terrorist and unconventional attacks on the territory of the continental United States, while U.S. forces moving through the global commons could find themselves under persistent and effective attack. In this respect, the immediate past is not necessarily a guide to the future.

Deterrence of aggression and of certain forms of warfare will remain an important element of U.S. national security strategy, and the fundamentals of deterrence theory will apply in the future as they have for thousands of years of human history. Deterrence operations will be profoundly affected by three aspects of the future joint operating environment.

First, U.S. deterrence strategy and operations will need to be tailored to **address multiple potential adversaries. A "one-size-fits-all" deterrence strategy** will not suffice in the future joint operating environment. Deterrence campaigns that are tailored to specific threats ensure that the unique decision calculus of individual adversaries is influenced.

Second, the increased role of transnational non-state actors in the future joint operating environment will mean that U.S. deterrence operations will have to find **innovative new approaches to "waging" deterrence against such adversaries.** Non-state actors differ from state actors in several key ways from a deterrence perspective. It is often more difficult to determine precisely who makes the key decisions one seeks to influence through deterrence operations. Non-state actors also tend to have different value structures and vulnerabilities. They often possess few critical physical assets to hold at risk, and are sometimes motivated by ideologies or theologies that make deterrence more difficult (though usually not impossible). Non-state actors are often dependent on the active and tacit support of state actors to support their operations. Finally, our future deterrence operations against non-state actors will likely suffer from a lack of well established means of communications that usually mark state-to-state relations.

Third, continued proliferation of weapons of mass destruction will make the U.S. increasingly the subject of the deterrence operations of others. As such, the U.S. may find itself in situations where its freedom of action is constrained unless it can checkmate the enemy's deterrent logic.

U.S. nuclear forces will continue to play a critical role in deterring, and possibly countering, threats to our vital interests in the future joint operating environment. Additionally, U.S. security interests will be advanced to the degree that its nuclear forces are seen as supporting global order and security. To this end, the U.S. must remain committed to its moral obligations and the rule of law among nations. It must provide an example of a responsible and ethical nuclear power in a world where nuclear technology is available to a wide array of actors. Only then will the existence of powerful U.S. nuclear forces, in support of the global order, provide friends and allies with the confidence that they need not pursue their own nuclear capabilities in the face of growing proliferation challenges around the world.

Unfortunately, we must also think the unthinkable – attacks on U.S. vital interests by implacable adversaries who refuse to be deterred could involve the use of nuclear weapons or other WMD. For both deterrence and defense purposes our future forces must be sufficiently diverse and operationally flexible to provide a wide range of options to respond. Our joint forces must also have the recognized capability to survive and fight in a WMD, including nuclear, environment. This capability is essential to both deterrence and effective combat operations in the future joint operating environment.

If there is reason for the joint force commander to consider the potential use of nuclear weapons by adversaries against U.S. forces, there is also the possibility that sometime in the future two other warring states might use nuclear weapons against each other. In the recent past, India and Pakistan have come close to armed conflict beyond the perennial skirmishing that occurs along their Kashmir frontier. Given India's immense conventional superiority, there is considerable reason to believe such a conflict could lead to nuclear exchanges. As would be true of any use of nuclear weapons, the result would be massive carnage, uncontrolled refugee flows, and social collapse -- all in all, a horrific human catastrophe. Given 24/7 news coverage, the introduction of U.S. and other international forces to mitigate the suffering would seem to be almost inevitable.

> While we continue to bin the various modes of war into neat and convenient categories, it should be recognized that future adversaries do not have the same lens or adhere to our Western conventions of war. In fact, there is a great amount of granularity across the spectrum of conflict, and a greater **potential for "hybrid" types of war.** This **assessment** acknowledges the blending of regular and irregular forms of warfare. It has also identified a convergence between some terrorist organizations and transnational crime. Some have postulated a further blurring of these various modes of conflict and challenges to governance as part of the future operating environment. To the historically minded, in fact, there is nothing new in such an approach. The Southern campaigns of the American Revolutionary War, the advanced European weapons and tactics exploited by the Boers at the turn of the 20th **century, and General William Slim's Burma** campaign provide evidence regarding the results that can be obtained by combining the diffuse nature of irregular methods with modern weaponry. Wars of the twenty-first century will similarly see no clear distinction between the methods used to achieve victory. Future opponents will exploit whatever methods, tactics, or technologies that they think will thwart us.

Nuclear and major regular war may represent the most important conflicts the Joint Force could confront, but they remain the least likely. Irregular wars are more likely, and winning such conflicts will prove just as important to the protection **of America's vital interests and the maintenance** of global stability.

A significant component of the future operating environment will be the presence of major actors which are not states. A number of transnational networked organizations have already emerged as threats to order across the globe. These parasitic networks exist because communications networks around the world enable such groups to recruit, train, organize, and connect. A common desire to transcend the local regional, and international order or challenge the traditional power of states characterizes their culture and politics. As such, established laws and conventions provide no barrier to their actions and activities. These organizations are also becoming increasingly sophisticated, well-connected, and well-armed. As they better integrate global media sophistication, lethal weaponry, potentially greater cultural awareness and intelligence, they will pose a considerably greater threat than at present. Moreover, unburdened by bureaucratic processes, transnational groups are already showing themselves to be highly adaptive and agile.

Irregular adversaries will **use the developed world's conventions and moral** inhibitions against them. On one hand the Joint Force is obligated to respect and adhere to internationally accepted "laws of war" and legally binding treaties to which the United States is a signatory. On the other hand, America's enemies, particularly the non-state actors, will not find themselves so constrained. In fact, they will likely use law and conventions against the U.S. and its partners.

That said, in the end irregular war remains subject to the same fundamental dynamics of all wars: political aims, friction, human frailties, and human passion. Nevertheless, the context within which they occur does contain substantial differences. As Mao suggested, the initial approach in irregular war must be a general unwillingness to engage the regular forces they confront. Rather, according to him, they should attack the enemy where he is weakest, and in most cases this involves striking his political and security structures. It is likely that the enemy will attack those individuals who represent the governing authority or who are important in the local economic structure: administrators; security officials; tribal leaders; school teachers; and business leaders among others, particularly those who are popular among the locals. If joint forces find themselves engaged in such situations, a deep understanding of the local culture and the political situation will be fundamental to success.

Source: The Intelligence & Terrorism Information Center

South Lebanon 2006: Hezbollah rockets Israel

What past irregular wars have suggested is that military organizations confronted by irregular enemies must understand the "other." Here, the issue is to understand not just of the nature of the conflict, but the "human sea," to use Mao's analogy, within which the enemy swims. The great difficulty U.S. forces will confront in facing irregular warfare is that such conflicts require a thorough understanding of the cultural, religious, political, and historical context within which they are being fought, as well as a substantial commitment of "boots on the ground" for sustained periods of time. There are no "rapid decisive operations" in irregular warfare that can achieve swift victory. Instead of decisive campaigns, U.S. forces can only achieve victory by patient, long-term commitments to a consistent, coherent strategic and political approach.

This coherent approach must also take into account the capabilities of other elements of government. Often, interagency cooperation is difficult because of the relative imbalance of resources between the Department of Defense and other agencies. For this reason, the Joint Force can expect tension to exist between tasks that must be completed to accomplish the mission, and enabling the interagency community to engage effectively. Ultimately, war against irregular enemies can only in the end be won by local security forces. Moreover, the indices of success are counter intuitive: fewer engagements, not more; fewer arms captured, not more; fewer enemy dead, not more.

What is of critical importance in irregular war is the ability to provide security to the local population with the purpose of denying the enemy the ability to survive among the people, allowing local police and military forces to build up sufficient strength to control their area of responsibility. Moreover, the Joint Force should contribute to the development of political legitimacy so that local police and military forces are acting with the support of the local population and not against it. The security side of the mission requires a deep understanding of local culture, politics, history, and language. In all cases the use of firepower will be a necessary feature, but balanced with non-lethal activities. Equally important will be the provision of high quality advisors to indigenous forces. Ultimately, U.S. forces can neither win a counterinsurgency, nor ensure that indigenous forces are regarded as the legitimate governing authority; only the locals can put in place the elements guaranteed to achieve lasting victory.

The current demographic trends and population shifts around the globe underline the increasing importance of cities. The urban landscape is steadily growing in complexity, while its streets and slums are filled with a youthful population that has few connections to their elders. The urban environment is subject to water scarcity, increasing pollution, soaring food and living costs, and labor markets, in which workers have little leverage or bargaining power. Such a mixture suggests a sure-fire recipe for trouble.

Thus, it is almost inevitable that joint forces will find themselves involved in combat or relief operations in cities. Such areas will provide adversaries with environments that will allow them to hide, mass, and disperse, while using the cover of innocent civilians to mask their operations. They will also be able to exploit the interconnections of urban terrain to launch attacks on infrastructure nodes with cascading political effects. Urban geography will provide enemies with a landscape of dense buildings, an intense information environment, and a complexity all of which makes defensive operations that much easier to conduct. The battles of Leningrad, Stalingrad, Seoul, and Hue with their extraordinarily heavy casualties all offer dark testimony to the wisdom of Sun Tzu's warning: "The worst policy is to attack cities. Attack cities only when there is no alternative."[27]

If there is no alternative than to fight in urban terrain, joint force commanders must prepare their forces for the conduct of prolonged operations involving the full range of military missions. They should do so cognizant that any urban military operation will require a large number of troops and that actual urban combat could consume manpower at a startling rate. Moreover, operations in urban terrain will confront joint force commanders with a number of conundrums. The very density of building and population will inhibit the use of kinetic means, given the potential for collateral damage as well as large numbers of civilian casualties. Such inhibitions could increase U.S. casualties. On the other hand, any collateral damage carries with it difficulties in winning the "battle of the narrative." How crucial the connection between collateral damage and disastrous political implications is suggested by the results of a remark an American officer made during the Tet offensive that American forces "had to destroy a village to save it." That comment reverberated throughout the United States and was one of the contributing factors to the erosion of political support for the war.

The ability of terrorists to learn from their predecessors and colleagues will not confront the hindrance of having to process adaptations and innovations through bureaucratic barriers. One must also note the growing convergence of terrorist organizations with criminal cartels like the drug trade to finance their activities. Such cooperative activities will only make terrorism and criminal cartels more dangerous and effective.

Operations against terrorists will keep Special Forces busy, with conventional forces increasingly active in supporting and complementary roles. If the Middle East continues on its troubled path, it is likely the war on terrorism will not continue on its current levels, but could actually worsen. Where an increase in terrorist activity intersects with energy supplies or weapons of mass destruction, joint force commanders will confront the need for immediate action, which may require employment of significant conventional capabilities.

Finally, we should underline that persistent media coverage, coupled with changing Western attitudes about the use of force, will influence and be influenced by U.S. military operations. What will be of great importance in the situations where force is being employed will be the narrative that plays on the world's stage. The joint force commander must understand that he should place particular emphasis on creating and influencing that narrative. Moreover, he must be alert and ready to counter the efforts of the enemies of the United States to create and communicate their own narratives. The enemy's ability to operate within the local cultural and social fabric will complicate such efforts. This puts at a premium the ability of Americans to understand the perceptual lenses through which others view the world.

Source: The Intelligence & Terrorism Information Center

High Tech Guerrilla command post: Hezbollah position South Lebanon 2006 with off-the-shelf sensors

D. Professional Military Education: The Critical Key to the Future

The future Chairmen of the Joint Chiefs of Staff of the 2030s and the Service Chiefs of Staff are already on active duty in the rank of Captain or Lieutenant. The Combatant Commanders and all the future flag and general officers of the U.S. military in the 2030s are currently on active duty. The Command Sergeants Major and Command Master Chiefs of the Joint Force in 2030 are in uniform. In other words, preparation of the senior military leaders of the 2030s has already begun!

As SirMichael Howard once commented, the military profession is not only the most demanding physically, but the most demanding intellectually. Moreover, it confronts a problem that no other profession possesses:

> *There are two great difficulties with which the professional soldier, sailor, or airman has to contend in equipping himself as commander. First, his profession is almost unique in that he may only have to exercise it once in his lifetime, if indeed that often. It is as if a surgeon had to practice throughout his life on dummies for one real operation; or a barrister only appeared once or twice in court towards the close of his career; or a professional swimmer had to spend his life practicing on dry land for the Olympic Championship on which the fortunes of his entire nation depended. Secondly the complex problem of running a [military service] at all is liable to occupy his mind so completely that it is easy to forget what it is being run for.*[28]

While the preparation of these young officers and NCO's must begin with their training as military professionals, it must also include their intellectual education to confront the challenges of war, change, and differing cultures. In the space of twenty-five years, they must master the extraordinarily difficult tasks of the irmilitary specialties as well as those required by joint warfare. But equally important, they must prepare themselves for the challenges presented by war and the projection of military force.

The recent experiences of Afghanistan and Iraq have made clear that in war, human beings matter more than any other factor. There are other dimensions, including technology, that are important, but rarely decisive. Above all, officers who hold the senior positions in the American military in the 2030s must develop a holistic grasp of their professional sphere and its relationship to strategy and policy. At this level of leadership, the skills for building trust that will serve as the foundation for harmonious teams is as important as tactical or operational prowess -maybe more so. The future Joint Force must have leaders who can form and lead effective coalitions. Such a preparation will take a lifetime of intellectual preparation, because it demands an ability to understand the "other" in his terms, historically, politically, culturally, and psychologically.

The world of the 2030s will demand more than mastery of the technical and operational aspects of war. The nature of the decentralized operations required by many of the challenges described thus far will require that NCOs must also understand the fundamental nature of war as well as other cultures and peoples – as they will undoubtedly confront challenges equivalent to those faced by today's mid-grade officer. Both officers and enlisted leaders will find themselves participating in coalitions, in which the United States may or may not be the leading actor, but in which partners will invariably play an important part. All military leaders must be equipped with the confidence to decide and act in ambiguous situations and under conditions where clear direction from above may be lacking or overcome by changing conditions.

This is the fundamental challenge the U.S. military will confront: providing the education so that future leaders canunder stand the political, strategic, historical, and cultural framework of a more complex world, as well as having a thorough grounding in the nature of war, past, present, and future. Admiral Stansfield Turner, initiator of the intellectual revolution at the Naval War College in the early 1970s, best expressed the larger purpose of professional military education:

> *War colleges are places to educate the senior officer corps in the larger military and strategic issues that confront America... They should educate these officers by a demanding intellectual curriculum to think in wider terms than their busy operational careers have thus far demanded. Above all the war colleges should broaden the intellectual and military horizons of the officers who attend, so that they have a conception of the larger strategic and operational issues that confront our military and our nation.*

The complexity of the future suggests that the education of senior officers must not remain limited to staff and war colleges, but should extend to the world's best graduate schools. Professional military education must impart the ability to think critically and creatively in both the conduct of military operations and acquisition and resource allocation. The services should draw from a breadth and depth of education in a range of relevant disciplines to include history, anthropology, economics, geopolitics, cultural studies, the 'hard' sciences, law, and strategic communications. Their best officers should attend such programs. Officers cannot master all these disciplines, but they can and must become familiar with their implications. In other words, the educational development of

America's future military leaders must not remain confined to the school house, but must involve self study and intellectual engagement by officers throughout their careers.

PART V: SOME LEADING QUESTIONS

Despite the uncertainties and ambiguities involved in the future security environment there are two specific areas where the U.S. military can better prepare its forces and its future leaders to meet the challenges that will come. As this study suggested at the beginning, perhaps the most important cultural attributes military organizations require are the ability to innovate in peacetime and adapt in war to the actual realities of the battlefield. Unfortunately the present culture and bureaucratic structures of the Department of Defense place major hurdles in the path of future innovation and adaptation.

One can encapsulate those obstacles in simple words or phrases. What needs reform is obvious, but the actual execution, **the important "how to," of any effective reform** will require sustained efforts against comfortable, deeply entrenched bureaucracies, sub-cultures within the military, and the demands of the present. Two areas that demand change are acquisition and the personnel systems.

A. Defense Economics and Acquisition Policies

The *Joint Operating Environment* has spoken thoroughly about the asymmetric application of power by potential enemies against U.S. military forces. There is also an asymmetry **with respect to the "defense spending" of the** United States and its potential opponents, particularly in irregular contexts. One only need to consider the enormous expenditures the United States has made to counter the threat posed by improvised explosive devices (IED). The United States has spent literally billions to counter these crude, inexpensive, and extraordinarily effective devices. If one were to multiply this ratio against a global enemy, it becomes unexecutable. While this asymmetry is most dramatic against the low-end threat, it applies to more sophisticated threats as well. Current economics indicate that China likely spends far less than the United States for the same capability. For instance, because of its labor market, the cost of many of the raw materials, and the savings gained by reverse engineering technologies, the

Chinese space program costs an order of magnitude less than that of the United States.

There have been justified calls for acquisition reform for decades, and while a number of groups have produced clear, forthright, and intelligent studies, little actual reform has taken place. This is no longer a bureaucratic issue – it is having strategic effects. Given the potential for disruptive technologies in the near future, the crucial issue will not be whether the United States possesses such technologies, but how affordably, how quickly, and how effectively joint forces can incorporate those technologies not only into their concepts, doctrine, and approach to war, but actually into the units and commands that will have to use those technologies on future battlefields.

Without a thorough and coherent reform of the acquisition processes, there is the considerable prospect an opponent could incorporate technological advances more affordably, quickly, and effectively – with serious implications for future joint forces.

B. The Personnel System

Perhaps the greatest difficulty confronting the Joint Force in preparing future leaders has to do with a personnel system that derives its philosophical and instrumental basis from reforms conducted between 1899 and 1904 and laws passed by Congress in 1947, 1954, and 1986. To a considerable degree, these reforms and laws still drive Service approaches to recruiting, training, promoting, and eventually retiring their personnel.

The current personnel and leader development system has its roots in long outdated mobilization systems for mass armies in world wars. And while the United States has had an all-volunteer force for 35 years, the bureaucracy still "thinks" and "acts" from an industrial-age, mobilization-based leader development paradigm. That approach continues to shape how the services approach training and education, often confusing the two. That state of affairs must change.

If we expect to develop and sustain a military that operates at a higher level of strategic and operational understanding, then the time has come to address the recruiting, education, training, incentive, and promotion systems so that they are consistent with the intellectual requirements for the future joint force.

PART VI: CONCLUDING THOUGHTS

Do make it clear that generalship, at least in my case, came of understanding, of hard study and brain-work and concentration. Had it come easy to me, I should not have done [command] so well. If your book could persuade some of our new soldiers to read and mark and learn things outside drill manuals and tactical diagrams, it would do a good work. I feel a fundamental crippling in curiousness about our officers. Too much body and too little head. The perfect general would know everything in heaven and earth.

So please, if you see me that way and agree with me, do use me as a text to preach for more study of books and history, a greater seriousness in military art. With two thousand years of example behind us, we have no excuse, when fighting, for not fighting well...[29]

T.E. Lawrence to B.H. Liddell Hart, 1933

The ability to innovate in peacetime and adapt during wars requires institutional and individual agility. This agility is the product of rigorous education, appropriate applications of technology and a rich understanding of the social and political context in which military operations are conducted. But above all, innovation and adaptation require imagination and the ability to ask the right questions. They represent two of the most important aspects of military effectiveness. The former occurs during peace, when there is time available to think through critical issues. However, in peacetime, military organizations cannot replicate the actual conditions of combat, when a human opponent is trying his best to destroy U.S. forces. Thus, there must be a premium on studying the military - from an evidence-based perspective, using history, current operations, wargames, and experiments -to better understand the present and future. There must be a connection between those in the schools and those involved in experimentation. Above all, there must be rigorous, honest red teaming and questioning of assumptions. "All the objectives were met" is a guarantee of intellectual dishonesty as well as a recipe for future military disaster.

Adaptation provides little time for reflection because of the immediate demands of combat. Here the patterns of thought developed in peacetime are crucial, because adaptation requires the questioning of the assumptions with which military organizations have entered the conflict. In the past, military organizations which have ruthlessly examined and honestly evaluated their assumptions in peacetime have done the same in war. Those which have not, have invariably paid a terrible price in lives. Those, whose commanders have listened and absorbed what their subordinates have had to say, were those which recognized what was actually happening in combat, because they had acculturated themselves to learning from the experiences of others.

The defining element in military effectiveness in war lies in the ability to recognize when prewar visions and understanding of war are wrong and must change. Unfortunately in terms of what history suggests, most military and political leaders have attempted to impose their vision of future war on the realities of the conflict in which they find themselves engaged, rather than adapting to the actual conditions they confront. The fog and friction that characterize the battle space invariably make the task of seeing, much less understanding what has actually happened, extraordinarily difficult. Moreover, the lessons of today, no matter how accurately recorded and then learned, may no longer prove relevant tomorrow. The enemy is human and will consequently learn and adapt as well. The challenges of the future demand leaders who possess rigorous intellectual understanding. Providing such grounding for the generals and admirals, sergeants and chiefs of the 2030s will ensure that the United States is as prepared as possible to meet the threats and seize the opportunities of the future.

End Notes

[1] Sun Tzu, *The Art of War*, trans. and ed. by Samuel B. Griffith (Oxford, 1963), p. 63.
[2] Thucydides, *The History of the Peloponnesian War*, trans. by Rex Warner (London: Penguin Books, 1954), p. 48.
[3] Quoted in Colin Gray, *Another Bloody Century*, (London: Penguin Books, 2005), p. 40.
[4] Thucydides, *History of the Peloponnesian War*, trans. by Rex Warner (London: Penguin Books, 1954) p. 80.
[5] Carl von Clausewitz, *On War*, translated and edited by Michael Howard and Peter Paret (Princeton, NJ: Princeton University Press, 1976), p. 87.
[6] Carl von Clausewitz, *On War*, translated and edited by Michael Howard and Peter Paret (Princeton, NJ: Princeton University Press, 1976), p. 113.
[7] Barry D. Watts, *Clausewitzian Friction and Future War* (Washington, DC: Institute for National Strategic Studies, 1992), pp. 122-123.
[8] Carl von Clausewitz, *On War*, translated and edited by Michael Howard and Peter Paret (Princeton, NJ: Princeton University Press, 1976) p. 77.
[9] Sun Tzu, *Art of War*, translation by Samuel B. Griffth (Oxford University), p.84.
[10] Winston S. Churchill, *The World Crisis* (Toronto: MacMillan, 1931), p. 6.
[11] Robert Heinl, *Dictionary of Military and Naval Quotations* (U.S.Naval Institute Press, 1976), p. 311.
[12] Sun Tzu, *The Art of War*, translated and edited by Samuel B. Griffith (Oxford, 1963), p. 92.
[13] **Center for Strategic International Studies (CSIS), "The Graying of the Great Powers,"** (Washington, DC), p. 7.
[14] Kerry Emanuel, Ragoth Sundaraarajan, **and** John Williams, *"Hurricanes and Global Warming,"* Bulletin American Meteorological Society, March 2008, pp. 347-367.
[15] Thucydides, *History of the Peloponnesian War,* trans. by Rex Warner (London: Penguin Books, 1954) p. 155.

[16] Colin Gray, "*Sovereignty of Context*", **Strategic Studies Institute, (2006)**
[17] Samuel Huntington, quoted in Joseph Nye (with Robert Keohane), *Power in the Global Information Age: From Realism to Globalization*. (London, 2004) p. 172.
[18] John Hamre, President, Center for Strategic and International Studies. (June 2007)
[19] The title of Colin Gray's book on the future of war in the twenty- first century.
[20] Quoted in Robert Debs Heinl, Jr., *Dictionary of Military and Naval Quotations* (Annapolis, MD, 1967), p.320.
[21] Christopher Pherson, "*Meeting the Challenge of China's Rising Power*," **Carlisle Papers in Security Strategy, July 2006.**
[22] Samuel Huntington, T*he Clash of Civilizations and the Remaking of World Order,* (New York: Simon and Schuster, 1996)
[23] R.V. Jones, *The Wizard War,* British Scientific Intelligence, 1939-1945 (New York, 1978), p. 199.
[24] Winston Churchill, *The Second World War, vol. 1, The Gathering Storm* (Boston: Houghton Mifflin, 1948), p. 156
[25] Sun Tzu, *The Art of War*, trans. and ed. by Samuel B. Griffith (Oxford, 1963), p.93.
[26] Quoted in Thomas Cleary, *The Japanese Art of War*, Understanding the Culture of Strategy (Boston, 1992), p. 38.
[27] Sun Tzu, *The Art of War*, As translated by Samuel B. Griffith (Oxford, 1963), p. 78.
[28] Sir Michael Howard, "The Uses and Abuses of Military History," Journal of the Royal United Service Institution, 107(1962), p. 6.
[29] As quoted in Robert B. Asprey, *War in the Shadows, The Guerrilla in History*, vol. 1 (Garden City, NY: Doubleday & Company, 1975), p. 270.

INDEX

A

accounting, 35
accuracy, 65
achievement, 28, 67
Acquired Immune Deficiency Syndrome, 68
acute, 20, 61
adaptability, 33
adaptation, vii, 5, 86, 88
adjustment, 31
administrative, 70
administrators, 80
Afghanistan, 10, 12, 42, 56, 60, 72, 75, 84
Africa, 8, 20, 21, 23, 27, 29, 32, 34, 35, 58, 59, 61, 68, 70
age, 8, 11, 17, 19, 87
aggression, 77
aggressive behavior, 26
agility, 7, 62, 88
agricultural, 35, 51
agriculture, 33, 35
aid, 58, 61
AIDS, 58, 68
air, 10, 59, 66, 67, 76
al Qaeda, 63
Alaska, 28
allies, 11, 33, 47, 54, 67, 72, 73, 74, 76, 78
alternative, 5, 27, 31, 82
alternative energy, 27, 31
ambiguity, 6

American Revolution, 79
anthropology, 85
anti-American, 44, 57
application, 86
appropriations, 54
Arabia, 27, 29
Argentina, 57
armed conflict, 78
armed forces, 8
Armenia, 53
Army, 11, 12, 47
Asia, 12, 23, 32, 38, 44, 45, 52, 55, 59, 61, 70, 75
Asian, 54, 55
assassination, 14
assault, 57, 62
assessment, 79
assets, 26, 32, 77
assumptions, 6, 9, 41, 51, 88
asymmetry, 86
Athens, 39
Atlantic, 12, 22, 76
atmosphere, 41
attacks, 5, 41, 56, 68, 75, 77, 78, 82
attitudes, 9, 16, 27, 83
authority, 14, 34, 80, 81
availability, 33
aviation, 71
awareness, 8, 9, 79
Azerbaijan, 53

B

bandwidth, 39
Bangladesh, 24
banking, 42
bargaining, 82
barriers, 63, 79, 82
basic services, 70
behavior, 13, 17, 26
Beijing, 50
beliefs, 6
benefits, 23, 41
binding, 80
biofuels, 28, 57
biological weapons, 64
Black Sea, 53
Boston, 90
bottleneck, 28
Brazil, 24, 28, 31, 44, 57
Brazilian, 57
Britain, 6, 11, 35, 66
buildings, 82
bureaucracy, 5, 87
Burma, 23, 79

C

cables, 22
calculus, 77
campaigns, 77, 79, 81
Canada, 21, 28, 35
candidates, 56
Caribbean, 18
cartel, 65
cartels, 57, 62, 82
Caspian, 28, 53
Caspian Sea, 28
cast, 45
casting, 15
catastrophes, 26, 37, 51
Caucasus, 52, 53, 59
cease-fire, 12
Census, 16, 20
Census Bureau, 16, 20
Central America, 57
Central Asia, 52, 59, 61
Chad, 36
chaos, 32, 37, 42, 62
Chechnya, 51
chemical weapons, 5
children, 19
Chile, 57
China, 10, 12, 16, 17, 19, 20, 21, 22, 24, 26, 29, 31, 37, 44, 45, 46, 47, 48, 49, 50, 52, 54, 55, 57, 60, 63, 67, 86, 90
CIA, 68
citizens, 21, 42, 68
civil war, 20, 46, 51, 61
civilian, 8, 51, 71, 82
classical, 47
closure, 55
coal, 27, 28, 31
coalitions, 45, 57, 84, 85
cocaine, 57
codes, 42
Cold War, 10, 15, 44, 47, 48, 56, 63, 68
collateral, 82
collateral damage, 82
colleges, 85
Colombia, 28, 57
colonial power, 58
combustion, 22
commerce, 22, 55
commodities, 59
commodity, 44
commons, vii, 5, 6, 44, 73, 76, 77
communication, 40
communities, 8, 10, 11, 21
community, 63, 68, 81
competition, 6, 35, 47
competitor, 47
complexity, 14, 61, 64, 70, 81, 82, 85
composition, 19
computer systems, 66
computing, 9, 39
concentration, 88
conception, 85
conceptualization, 73
confidence, 10, 69, 78, 85

Index

conflict, 4, 5, 6, 8, 9, 10, 12, 21, 23, 31, 35, 36, 41, 42, 53, 54, 56, 59, 60, 61, 62, 72, 73, 74, 77, 78, 79, 81, 88, 89
confrontation, 10, 47
confusion, 8, 70
Congress, 39, 87
consensus, 47
conservation, 27, 31
constraints, 28, 62, 75
construction, 29
control, 14, 22, 38, 39, 58, 60, 61, 66, 68, 70, 71, 75, 76, 77, 81
convergence, 79, 82
corn, 34
corruption, 53
costs, 15, 17, 62, 64, 81, 87
countermeasures, 62
creativity, vii, 5, 7
creditor nation, 12
crime, 53, 79
criminal gangs, 51, 57, 62
criticism, 4
crops, 34
Cuba, 57
culture, 9, 10, 18, 21, 23, 42, 51, 63, 79, 80, 81, 86
currency, 23
curriculum, 85

D

danger, 23, 61, 76
Darfur region, 36
death, 4, 77
deaths, 17, 23, 38
debtor nation, 12
decisions, 77
defense, 11, 33, 41, 47, 48, 66, 67, 72, 74, 75, 78, 86
delivery, 54, 63, 75
democracy, 3, 8
demographic transition, 53
demographics, 16
Deng Xiaoping, 45
density, 68, 82

Department of Defense, 12, 46, 58, 81, 86
depression, 13
destruction, 5, 60, 64
deterrence, 54, 63, 73, 74, 77, 78
detonation, 66
developed countries, 16, 20
developed nations, 35
developing countries, 16, 17
developing nations, 27, 35
diet, 20
dietary, 33
diffusion, 62
disaster, 68, 69, 71, 75, 88
disaster relief, 69
discounting, 40
diseases, 70
dishonesty, 88
dislocations, 21, 39
disorder, 34, 72
distress, 37
dividends, 68
doctors, 21, 38
domestic economy, 27
dominance, 41, 46, 55, 60, 73
dumping, 37

E

earth, 88
East Asia, 38, 44, 59, 75
East Timor, 51
Eastern Europe, 56
economic activity, 53
economic change, 73
economic crisis, 26
economic development, 46
economic growth, 17, 20, 24, 26, 44, 49, 55, 58
economic integration, 73
economic stability, 27, 61, 72
economic systems, 15
economics, 21, 26, 38, 42, 85, 86
ecosystems, 37
Education, 84
educational system, 67

Egypt, 44
El Salvador, 10
elders, 81
electricity, 28
e-mail, 69
EMP, 66
employment, vii, 4, 5, 15, 20, 24, 69, 73, 83
empowered, 62, 65
energy, 12, 13, 27, 28, 31, 33, 44, 48, 60, 63, 71, 83
engagement, vii, 5, 24, 58, 60, 61, 86
environment, 1, 3, 6, 8, 26, 40, 41, 42, 45, 52, 60, 69, 73, 77, 78, 79, 81, 82, 86
ethnicity, 56
Europe, 6, 12, 20, 21, 23, 38, 45, 55, 56, 68, 75
European Union, 44, 55
Europeans, 6, 22, 44, 56
evil, 68
excuse, 88
execution, 8, 15, 86
exercise, 55, 84
expanded trade, 49
expenditures, 48, 74, 86
expertise, 58
exports, 59
extreme poverty, 24
extremism, 44, 56

F

fabricate, 64
failed states, 27
failure, 9, 12, 23, 65, 66, 68, 75
family, 17
famine, 34
fanaticism, 60
fear, 8, 9, 28, 38, 52
federal government, 68
feedback, 62
females, 17, 19
finance, 82
financial crises, 26
financial system, 26, 59
fire, 12, 28, 71, 82

First World, 23
fissile material, 63
flight, 66
flow of capital, 22
focusing, 12
food, 33, 34, 70, 81
foodstuffs, 28
forecasting, 12, 16
foreign exchange, 51
fractures, 26
France, 11
freedom, 78
friction, 9, 39, 80, 89
fuel, 44, 75
funds, 32

G

gas, 27, 31, 51, 56
GDP, 22, 24, 44, 48
GDP per capita, 44
generation, 61
Geneva, 42
Geneva Convention, 42
genocide, 8
genome, 13
geography, 9, 23, 42, 82
Georgia, 50, 53, 68
German philosophy, 67
Germany, 11, 44, 46, 75
global communications, 69
global economy, 17, 22, 26, 55
Global Positioning System, 11
global resources, 27
global trade, 23
Global Warming, 37, 89
globalization, 12, 21, 22, 23, 42, 73
Globalization, 21, 22, 90
goals, 33, 44, 73
governance, 58, 79
government, 1, 17, 22, 26, 46, 50, 51, 57, 62, 68, 69, 74, 81
government policy, 1
GPS, 11
graduate education, 67

Great Depression, 11, 26, 27, 32, 33
Great War, 14
Greece, 7
green revolution, 33
Grenada, 11
grounding, 85, 89
groundwater, 35
groups, 8, 36, 42, 44, 58, 60, 61, 62, 63, 64, 65, 68, 79, 87
growth, 16, 17, 18, 19, 20, 24, 26, 27, 29, 31, 34, 43, 44, 49, 51, 55, 57, 58
growth rate, 16, 27

H

Haiti, 51
handling, 24, 67
hands, 33, 76
harm, 56
harvest, 35
health, 6, 17, 37, 39, 58, 69
heart, 73
height, 66
helplessness, 8
Hezbollah, 62, 80, 83
hibernation, 74
Hispanic, 18
Hispanic population, 18
holistic, 66, 84
holistic approach, 66
homeland security, 57, 62
horizon, 44
host, 14
hostage, 76
hostile environment, 40
human, 4, 5, 6, 7, 8, 9, 10, 13, 14, 21, 23, 26, 37, 42, 49, 61, 73, 77, 78, 80, 81, 84, 88, 89
human activity, 8
human condition, 6, 9
human nature, 5, 7, 23
humanitarian, 58, 71
humanity, 21, 22
Hurricane Andrew, 68
Hurricane Katrina, 68
hurricanes, 37
hybrid, 79

I

id, 11, 22, 49
ideology, 10
illusion, 4, 23, 60
imagination, 7, 65, 66, 88
immigrants, 18
immigration, 18, 44, 71
imports, 48, 70
in situ, 78
incentive, 59, 63, 87
incentives, 31
income, 33
increased competition, 41
India, 10, 19, 20, 21, 24, 31, 44, 54, 55, 63, 67, 78
Indian Ocean, 54, 55
indicators, 5, 12, 16, 41
indices, 81
indigenous, 59, 81
Indonesia, 24, 50
industrialization, 37, 59, 87
industry, 35
inequality, 21
inferiority, 54
infinite, 9
influenza, 38
Information Age, 90
information technology, 40
infrastructure, 31, 44, 49, 51, 57, 62, 70, 82
innovation, 86, 88
insight, 7
inspiration, 4, 45
instabilities, 73
instability, 32, 53, 59, 63
institutions, 26, 42, 66
integration, 49, 61, 73
intelligence, 10, 46, 67, 69, 77, 79
interactions, 14, 21
interdependence, 59
interference, 58
internal combustion, 22
international markets, 12

international trade, 22, 42
internet, 12, 21, 59, 60, 63, 67
intervention, 24, 27
intimidation, 4, 45
investment, 28, 31, 44, 53, 58
iPod, 39
Iran, 12, 23, 57, 63
Iraq, 8, 10, 23, 36, 42, 60, 72, 75, 76, 84
Iron Curtain, 72
irrigation, 35
Islam, 6, 33, 59
Islamic, 44, 56, 59, 60
Islamic world, 59, 60
isolation, 46
Israel, 36, 63, 80
Italy, 11

J

Japan, 11, 12, 17, 20, 23, 26, 32, 35, 54, 55, 63, 75
Japanese, 11, 17, 46, 50, 74, 90
jobs, 24
Joint Chiefs, 84
Jordan, 36
Jordanian, 36
judiciary, 42

K

Kashmir, 54, 55, 78
Katrina, 68
key indicators, 41
Korea, 1, 11, 35, 54
Korean, 12, 64

L

labor, 21, 82, 86
land, 22, 34, 35, 84
landscapes, 12
language, 81
Latin America, 21, 57
law, 28, 39, 78, 80, 85

laws, 79, 80, 87
leadership, 24, 50, 51, 61, 84
learning, 9, 10, 73, 89
Lebanon, 62, 80, 83
life expectancy, 51
life-threatening, 38
lifetime, 84
lift, 34, 39
likelihood, 9, 17, 61, 77
limitation, 11, 17
limitations, 15, 33
Lincoln, 14
linear, 16, 40
linguistic, 58, 61
liquidity, 26
local authorities, 68
location, 55, 56, 74
logistics, 34, 75, 76
London, 89, 90
losses, 17
low-intensity, 4

M

magnetic, 66
magnets, 67
mainstream, 18
maintenance, 73, 74, 79
major cities, 70
males, 17, 19
management, 61
man-made, 16, 51
manpower, 65, 82
Marine Corps, 4
maritime, 47
market, 31, 65, 86
markets, 12, 22, 76, 82
Marxist, 10
mask, 82
mass media, 67
mastery, 85
meals, 68
measures, 26, 27, 31, 39, 48, 69
media, 23, 24, 59, 67, 68, 69, 71, 79, 83
medical services, 17

men, 8, 14, 33, 39
Mexican, 57, 62
Mexico, 18, 24, 57, 61, 62
microbes, 38
middle class, 20
Middle East, 10, 20, 21, 27, 38, 55, 59, 60, 61, 70, 72, 82
migration, 22, 42
military, vii, 4, 5, 6, 7, 8, 9, 10, 11, 12, 15, 17, 20, 23, 24, 27, 32, 33, 37, 39, 41, 44, 45, 46, 47, 48, 49, 51, 54, 55, 56, 57, 60, 62, 65, 66, 68, 69, 71, 72, 73, 74, 75, 76, 81, 82, 83, 84, 85, 86, 87, 88, 89
military spending, 48
militias, 62, 65
minorities, 50, 54
missiles, 76
missions, 33, 39, 60, 72, 73, 76, 82
modernity, 59
modernization, 55, 59
Moldova, 53
money, 21
monopoly, 62
Morocco, 59
Moscow, 51
motives, 8
movement, 11, 22, 60, 66
multiplicity, 20
Muslims, 20, 21

N

nanotechnology, 12
narratives, 69, 83
NASA, 60
nation, 4, 5, 12, 18, 20, 22, 29, 47, 48, 52, 65, 68, 69, 72, 84, 85
National Guard, 68
national interests, 6, 33, 72, 77
national security, 1, 28, 55, 57, 77
National Strategy, 39
nationalism, 54
NATO, 12, 54, 55, 56
natural, 15, 16, 26, 31, 33, 34, 35, 37, 38, 51, 58, 76

natural disasters, 15, 34, 37, 76
natural gas, 31, 51
natural resources, 51, 58
Navy, 47
Nazi Germany, 23, 49, 54, 66
Near East, 35
network, 40, 67
new media, 69
New York, 14, 90
news coverage, 78
newspapers, 68
Nigeria, 19, 24, 44
Nile, 46
non-nuclear, 66
norms, 9, 42, 62
North Africa, 32, 35, 61
North America, 75
North Atlantic, 12
North Atlantic Treaty Organization, 12
North Korea, 1, 11, 23, 54, 63, 64
nuclear, 1, 5, 10, 11, 27, 28, 47, 54, 55, 61, 63, 64, 66, 74, 75, 78
nuclear power, 1, 29, 54, 63, 64, 78
nuclear technology, 78
nuclear threat, 75
nuclear weapons, 10, 11, 61, 63, 74, 78

O

obligations, 78
obsolete, 23
oceans, 37, 75
OECD, 27
off-the-shelf, 83
oil, 27, 28, 29, 30, 31, 32, 44, 48, 51, 52, 53, 55, 57, 65
oil production, 31
oil recovery, 30
oil revenues, 57
oil shale, 30
OPEC, 28, 29, 32, 33
open markets, 8
openness, 4
opposition, 28
optimism, 22, 29, 31

Organization for Economic Cooperation and Development, 27
organized crime, 53
overload, 9

P

Pacific, 8, 11, 37, 45, 54
pain, 8
Pakistan, 24, 34, 54, 55, 59, 61, 63, 78
pandemic, 17, 38, 39
paranoia, 54
partnerships, 45
passports, 22
pathogens, 37, 38
per capita, 44, 48
perception, 67, 68
perceptions, 9, 10, 37, 40, 45, 68
Peru, 57
petroleum, 27, 31, 51
Philippines, 21, 24, 55
philosophical, 59, 87
philosophy, 67
pipelines, 29
plague, 39, 57
planning, 11, 13, 74, 75
plants, 28
platforms, 27, 31
play, 4, 15, 17, 18, 24, 28, 51, 56, 62, 72, 78, 85
point of origin, 38
poison, 54
police, 11, 62, 81
policy makers, 4, 15
political leaders, 9, 43, 63, 89
political legitimacy, 81
political power, 57
political stability, 6
politicians, 22, 62
politics, 17, 18, 21, 25, 42, 57, 73, 79, 81
pollution, 37, 50, 81
poor, 20, 21, 55
population, 17, 18, 19, 20, 21, 23, 24, 28, 29, 33, 35, 36, 38, 40, 50, 51, 53, 55, 70, 81, 82
population density, 70

population growth, 17, 19, 20, 33
population pyramid, 19
porous borders, 60
posture, 27, 72
poverty, 10, 12, 17, 20, 24, 55, 59, 61
power, 1, 7, 8, 9, 10, 11, 24, 26, 27, 29, 32, 40, 42, 44, 45, 46, 47, 48, 49, 51, 54, 57, 58, 59, 62, 63, 65, 68, 69, 72, 73, 75, 76, 78, 79, 82, 86
power plants, 29
powers, 1, 4, 9, 24, 42, 44, 45, 49, 54, 55, 57, 58, 61, 62, 63, 64, 66, 73, 76
prediction, 14
pre-existing, 70
premium, 71, 83, 88
president, 14
President Vladimir Putin, 51
pressure, 33, 58, 62
prices, 29, 31, 32
prisoners, 14
proactive, 26
producers, 31, 69
production, 27, 28, 31, 32, 33, 35, 51, 64
profit, 72
program, 63, 68, 87
proliferation, 41, 63, 74, 76, 78
prosperity, 23, 32, 33, 35, 49
protection, 35, 73, 79
public, vii, 17, 28, 38, 39, 65, 69
public opinion, 69
public policy, 17
pulse, 66
Putin, 51, 54

Q

quarantine, 38
questioning, 88

R

race, 11, 37, 46
racism, 68
radar, 66, 67
radio, 66

radius, 67
rain, 35
rainfall, 34
range, 5, 43, 44, 53, 62, 65, 67, 72, 75, 76, 78, 82, 85
rationality, 23
raw materials, 86
reading, vii, 5
reality, 6, 7, 9, 39, 43
recession, 13, 26, 33
recognition, 56
reconstruction, vii, 5, 60, 69, 71
recovery, 29, 30, 51
recruiting, 87
redundancy, 76
refining, 27, 31
reflection, 68, 88
reforms, 87
refugee flows, 78
regional, 24, 44, 53, 57, 64, 79
regular, vii, 5, 60, 73, 74, 75, 77, 79, 80
regulation, 22
relationship, 15, 45, 72, 84
reliability, 12
religion, 38, 39
religions, 20, 55, 59
repair, 28, 51, 75
reputation, 68
research and development, 11, 17
resentment, 21, 59
reserves, 27, 30
resolution, 26
resource allocation, 85
resources, 22, 23, 24, 27, 29, 33, 35, 51, 57, 58, 62, 63, 81
respiratory, 38
retaliation, 12
reunification, 50
rhetoric, 59
rhythms, 74
risk, 23, 37, 38, 77
rivers, 37
robotics, 12, 13, 17, 65
rocky, 35
Roman Empire, 55

rule of law, 78
Russia, 17, 19, 24, 29, 44, 51, 52, 53, 54, 55, 56, 57, 63, 68
Russian, 17, 50, 51, 53, 54, 56
Rwanda, 21

S

Saddam Hussein, 10, 23, 75
safety, 28, 69
salt, 4
Sarajevo, 61, 70
SARS, 38
satellite, 41, 65
Saudi Arabia, 27, 29
savings, 86
scarce resources, 29
scarcity, 35, 36, 37, 81
school, 47, 80, 85, 88
sea level, 37
Second World War, 47, 49, 67, 75, 76, 90
secondary education, 67
secular, 8
security, 1, 3, 4, 8, 23, 28, 34, 39, 41, 42, 51, 55, 57, 58, 59, 62, 63, 71, 77, 78, 80, 81, 86
security services, 51
seizure, 54
seller, 22
sensors, 83
series, 26, 46, 54, 66, 73
services, 17, 37, 51, 62, 70, 85, 87
sewage, 37
shape, 87
shocks, 4
short period, 55
shortage, 27
shortages, 31, 33, 34, 50
Siberia, 21, 50, 52
Sierra Leone, 42
signs, 54
singularities, 15
sites, 66
skills, 24, 84
slums, 37, 70, 81
social attitudes, 17

social fabric, 83
social ills, 26
social network, 37
social structure, 62
Somalia, 21, 27, 42
South America, 57
South Asia, 55, 59
South Korea, 11, 54
Southeast Asia, 12, 32
sovereignty, 45, 53, 62
Soviet Union, 10, 11, 12, 46, 48, 51, 54, 63
Spain, 35, 59
specter, 37
spectrum, 77, 79
speech, 6
speed, 65, 66, 67
spin, 68
sponsor, 74
stability, 6, 21, 24, 26, 27, 50, 52, 58, 61, 62, 64, 70, 72, 79
stabilize, 44
stable states, 5
stages, 12
starvation, 34
state control, 62
storms, 37
strain, 34
strength, 4, 23, 24, 44, 45, 72, 81
stress, 44, 70
stretching, 59
strikes, 64
students, 47, 67
submarines, 46, 47, 57
Sub-Saharan Africa, 20, 21, 58, 61
subsistence, 34
Sudan, 21, 29, 36
suffering, 20, 51, 61, 78
suicide bombers, 5, 8
summer, 66
Sun, 4, 16, 47, 72, 82, 89, 90
superiority, 78
superpower, 43, 54, 57
supply, 22, 28, 33, 34, 35
surplus, 31
surprise, 4, 6, 7, 15, 61, 65, 66, 73

surveillance, 77
survival, 4, 8
sustainability, 17
symbiotic, 72
synchronization, 62
Syria, 36

T

tactics, 63, 65, 79
Taiwan, 50, 54, 63
tanks, 50, 65
tar sands, 28, 31
targets, 64, 69
tariffs, 13
teachers, 80
technicians, 49
technological change, 40, 65
telephone, 21
television, 46
tension, 8, 15, 53, 55, 81
territory, 22, 54, 77
terrorism, 40, 56, 59, 62, 82
terrorist, 33, 38, 56, 57, 63, 68, 74, 77, 79, 82, 83
terrorist attack, 56, 68
terrorist organization, 38, 57, 63, 74, 79, 82
terrorists, 38, 52, 63, 67, 82
testimony, 42, 82
think critically, 85
thinking, 2, 6, 10, 13, 45, 47
third order, 72
Third Reich, 68
threat, 5, 11, 12, 44, 47, 50, 56, 57, 63, 66, 79, 86
threatening, 34, 38
threats, vii, 4, 5, 11, 21, 38, 75, 77, 78, 79, 86, 89
threshold, 54
Tibet, 18, 50
time frame, 19, 53
title, 90
tornadoes, 37
totalitarian, 32
trade, 12, 13, 22, 23, 24, 42, 49, 82

trading, 22, 23
training, 6, 11, 63, 74, 75, 84, 87
trajectory, 8, 13, 16, 24, 46
trans, 62, 89, 90
transfer, 12
translation, 89
transmission, 39, 69
transnational, 8, 44, 45, 59, 62, 77, 79
transportation, 28
travel, 22, 59
treaties, 11, 80
tribal, 36, 58, 59, 61, 80
trust, 69, 84
tsunamis, 37
Turkey, 36, 44

U

U.S. military, 6, 37, 45, 47, 74, 75, 83, 84, 85, 86
uncertainty, 4, 6, 61
unemployment, 27
uniform, 84
United Nations, 18, 42
United States, v, vii, 3, 8, 10, 11, 12, 15, 18, 19, 20, 21, 22, 23, 24, 27, 28, 29, 33, 36, 37, 38, 39, 40, 41, 42, 44, 45, 47, 48, 54, 55, 57, 63, 65, 66, 67, 68, 69, 72, 73, 74, 75, 76, 77, 80, 82, 83, 85, 86, 87, 89
universe, 10
unpredictability, 9
urban areas, 37, 70

V

values, 4, 8, 59
variables, 43
vehicles, 29
Venezuela, 29, 57
vessels, 35
victims, 37
Vietnam, 12, 24, 44, 55, 75
Vietnam War, 75
Vietnamese, 54
village, 82

violence, 17, 21, 23, 59, 62
violent, 17, 56, 61
virulence, 38
vision, 4, 15, 45, 89

W

war, 3, 4, 5, 6, 7, 8, 9, 10, 11, 12, 20, 21, 22, 23, 25, 32, 39, 42, 46, 47, 60, 61, 66, 69, 72, 73, 74, 75, 77, 79, 80, 81, 82, 84, 85, 86, 87, 88, 89, 90
war on terror, 82
warfare, 4, 47, 63, 77, 79, 81, 84
warlords, 46
Warsaw, 12
Warsaw Pact, 12
water, 35, 36, 37, 50, 59, 71, 81
weakness, 72
wealth, 22, 23, 32, 41, 55, 57, 58
weapons, 5, 8, 32, 41, 44, 59, 61, 63, 64, 65, 66, 69, 74, 76, 78, 79, 83
weapons of mass destruction, 64, 65, 78, 83
weapons of mass destruction (WMD), 65
web, 22
welfare, 17, 19, 37, 51
welfare system, 19
Western Europe, 44, 56
Western Hemisphere, 57
wheat, 34
winning, 15, 47, 60, 72, 73, 79, 82
wisdom, 4, 74, 82
WMD, 65, 78
women, 8
workers, 20, 21, 82
working population, 20
World War, 8, 11, 14, 23, 47, 49, 51, 55, 66, 67, 70, 75, 76, 90

Y

Yemen, 19
yield, 35
young men, 8, 33
Yugoslavia, 61